URBAN ETHNICITY IN THE UNITED STATES

Volume 29, URBAN AFFAIRS ANNUAL REVIEWS

URBAN ETHNICITY IN THE UNITED STATES

New Immigrants and Old Minorities

Edited by

LIONEL MALDONADO

and

JOAN MOORE

Published in cooperation with the Urban Research Center,
University of Wisconsin—Milwaukee

Volume 29, URBAN AFFAIRS ANNUAL REVIEWS

SAGE PUBLICATIONS
Beverly Hills   London   New Delhi

For information address:

SAGE Publications, Inc.
275 South Beverly Drive
Beverly Hills, California 90212

SAGE Publications India Pvt. Ltd.
M-32 Market
Greater Kailash I
New Delhi 110 048 India

SAGE Publications Ltd
28 Banner Street
London EC1Y 8QE
England

Printed in the United States of America

Library of Congress Cataloging in Publication Data

Main entry under title:

Urban ethnicity in the United States.

(Urban affairs annual reviews ; v. 29)
"Published in cooperation with the Urban Research Center, University of Wisconsin—Milwaukee."
Bibliography: p.
1. United States—Emigration and immigration—Addresses, essays, lectures.   2. Cities and towns—United States—Addresses, essays, lectures.   I. Maldonado, Lionel A., 1939-   II. Moore, Joan W.   III. University of Wisconsin—Milwaukee. Urban Research Center.   IV. Series.
HT108.U7   vol. 29 [JV6455]   307.7'6s   85-1824
ISBN 0-8039-2269-8   [304.8'73]
ISBN 0-8039-2270-1 (pbk.)

FIRST PRINTING

Contents

Prefatory Note:
The New American Ethnology
Scott Greer   7

Introduction
Lionel Maldonado and Joan Moore   13

**PART I: HISTORICAL, DEMOGRAPHIC, AND
ECOLOGICAL CONSIDERATIONS**

1. Immigration Reform and Immigration History:
   Why Simpson-Mazzoli Did Not Pass
   Margo Conk   25

2. Post-1965 Immigrants: Demographic and
   Socioeconomic Profile
   Morrison G. Wong   51

3. Immigration Issues in Urban Ecology:
   The Case of Los Angeles
   Philip Garcia   73

4. Improving the Data:
   A Research Strategy for New Immigrants
   Jose Hernández   101

PART II: INSTITUTIONAL CONTEXTS AND RESPONSES

5. Race, Color, and Language in
   the Changing Public Schools
   Ricardo R. Fernández and William Vélez   123

6. Urban Labor Markets and Ethnicity:
   Segments and Shelters Reexamined
   Marcia Freedman   145

7. Political Economy and the
   Social Control of Ethnic Crime
   Eleanor M. Miller and Lynne H. Kleinman   167

8. Ethnicity and Social Welfare in
   American Cities: A Historical View
   Robert S. Magill   185

9. Ethnicity, Mental Health, and
   the Urban Delivery System
   William T. Liu and Elena S.H. Yu   211

10. Rainbow's End: From the Old to the
    New Urban Ethnic Politics
    Steven P. Erie   249

    References   277
    About the Contributors   302

Prefatory Note:
The New American Ethnicity

☐ MUCH OF THE VALUE of social science lies in its ability to criticize and reformulate conventional wisdom and customary definitions. Thus the sociologist as debunker is frequently more valuable to his or her audience than is the sociologist as constructor of social fact. For out of the residue of past situations, long vanished, the wild guesses of religion, folklore, and primitive science, humankind accumulates an assortment of assumptions in its philosophy that heaven and earth have never dreamed of.

In the same way, sociology itself is prone to concretize shrewd insights, quantitative guesses, explanations (all of which made sense in other times, other places) and erect them into unexamined assumptions—that is, dogma. Ethnocentrism in place or time of observation haunts our theory, and thinking about ethnicity seems a serious case of this habit.

Thinking on ethnic relations in American sociology was largely carried out in the twentieth century, and was largely a response to the happenings of the early years, particularly the closing of the borders to immigrants in 1924. The authors of the chapters in this volume note some of the conditions that gave rise to the immigration quotas, but central perhaps was World War II and the massive increase of U.S. productivity in the virtual absence of migration (other than from Mexico). I speak of this as central because it signaled changes in production, resulting from the substitution of machines and nonhuman energy for human labor. The declining need for labor continued, masked by the Great Depression, for productivity per labor unit continued to increase during the 1930s. Thus one of the major reasons for immigration, demand for labor, declined while sentiment for exclusionary laws grew.

After the closing of the borders, it was possible to assume that American society was biologically a closed system, with existing groups, however selected, the basis for future populations. (This explains the panic among the demographically inclined at the falling fertility rates of "native whites" during the 1920s and 1930s.) Social scientists, always on the lookout for clarification via simplification, tended to assume a closed system of American population. This meant that the conceptualization of ethnicity and its career through time took a form something like that described below.

First, there were a limited number of groups involved, defined by their origins. There were Europeans, North and West or South and East. There were Blacks from Africa in the plantation South, thence in increasing numbers to cities in the North and South and West. There were Hispanics from Mexico, mostly new immigrants of the twentieth century. There was a small, concentrated, highly visible population of Jews from Eastern Europe, mostly urban, and smaller populations of Japanese, Chinese, and Indians.

Second, the framework used to conceptualize ethnicity was based upon cultural and social *processes*. It saw the typical career of a new population from a different society as one of social and spatial concentration and segregation, a process of acculturation, and finally assimilation. The process of acculturation occurred through the cultural engines of the larger society on the one hand (mass education, mass media, mass politics) and growing economic competence resulting in the "social anchors" of economic production, homeownership, and political loyalties on the other. The greatest loyalty from this point of view was to the new society, culture, and state, mediated by these acculturating social institutions.

Assimilation was, then, the eschatological point to the whole process. It was a continual process of recreating "the union." Thus it had a double import: It was an empirical theory of the melding of new recruits into a society, and it was a normative theory of the necessity and desirability of this outcome. Research and analysis tended, then, to concentrate on the conditions under which this process occurred faster, slower, unevenly, or not at all. It emphasized the "handicaps" of the immigrants—unfamiliarity with the American environment, often exaggerated by a rural background; a country of origin often living under despotism; and, above all, a tenacious culture brought from the homeland, usually organized around a religion and its practices— different from those current in America at the time. Poverty was, of course, a given.

The second focus of attention was upon the societal response to the new members of the society. Societal acceptance seemed to hinge on similarity of culture and adept behavior in urban, industrializing, democratic society. Those of higher social rank, Caucasoid and Protestant and from urban societies that had been longer in the country, were quickest to assimilate. There was, however, a great barrier signalized by the process of segregation and most evident in urban settlements. This was the stigma of foreignness, often identified and exaggerated by physical appearance; it is an aspect of the broader division noted in Max Weber's general concept of "status honor." The low status of the "hyphenated" Americans and those of sharply different phenotypes (especially Blacks, Asians, and Mexicans with strong Indian features) was reflected in the key social relations.

These included the relative deference, in both formal and informal interaction, accorded assimilated native and immigrant. They included social distance signified by living quarters and associated facilities, resulting in ethnic neighborhoods and ethnic jobs. Relative-status honor thus precluded access to equal-status primary relations between older Americans and new, effectively limiting intermarriage and the bonding that results from it. The indicators of status honor were also dynamics that tended to perpetuate a given hierarchy and, the older the original immigration, the closer to the English-speaking founders, the higher the status honor tended to be.

But the social differences thus signalized tended to give way to the opportunities of a rich, mobile, and growing society. Status honor was not unitary and the market provided ways to equalize status, or at least thoroughly confuse the issue. Politics in some degree had the same effect, and eventually the educational system and the media became dynamics as well as simple indicators or constituent parts of the hierarchy. As the new immigrants learned and achieved, they moved toward assimilation in appearance and fact, social distance declined and finally dissolved, and irrelevant bases for social judgment withered away. Thus the empirical theory, thus the normative vision.

Underlying this view of immigration, ethnic populations, and the process of their incorporation were several assumptions. The first was a closed population, and the second was an economy that would generate a demand for labor proportionate to the supply, and a quantum of surplus supporting the population at a stable and indeed improving level of material consumption. The United States, from its origins "land rich and labor poor," had been able to promise a future to newcomers in terms of political freedom and material improvement much greater than could the homelands.

The temporary nature of the system described and analyzed so optimistically in the first half of the century became evident when these assumptions were falsified by massive change. In 1965, during a period of full employment and steadily increasing real income, the immigration laws were so changed as to allow a great increase in the numbers of immigrants, a change in their origins, and changes in the criteria for entering the country. As a result the new groups included a radically different population, with heavy representation from Southeast Asia, the Hispanic countries of the southern Western Hemisphere, and such Mediterranean countries as Italy. Further, groups from certain countries had a radically different social rank from nineteenth century immigrants. Those from certain Asian societies and from Cuba and other South American nations included a high proportion of professional and other well-educated, highly skilled people.

Meanwhile, the application of nonhuman energy and machines to the world of work steadily increased production per worker, or, in a more somber key, made human labor redundant. This volume includes valuable detailed analyses of the process and its effects on the labor market. Several generalizations stand out: The hardest hit labor forces are those in the formerly labor-hungry areas of mass manufacturing and processing. These jobs had the characteristics of relatively low entering qualifications, secure employment, and high pay. They were the key area for the tendency called by some the "middle classicization of the working class." At the same time the comparable jobs, with low qualifications and easy entry, are now in service occupations and have the opposite characteristics—relative insecurity and low pay.

At the same time, the increasing scale of human interdependence was becoming global. The same changes in technology that improved machines and eliminated labor, that brought hundreds of thousands of new immigrants from Cuba or Vietnam to Florida or California in a matter of hours, allowed for the export of industry to countries such as Haiti, where cheap labor created strong competition on the American market with locally produced goods. Thus the effective labor market extended far beyond the limited population assumed in the older view of ethnicity and assimilation.

The strength of a working population is dependent upon its bargaining power. This is finally a matter of its purchase on economic function, its indispensability. This usually takes one of two forms—

either the criticality of high skills or the need for large, dependable, disciplined labor supply. The first produced craft unions; the second, industrial unions. It is clear that each is threatened, the first by obsolescence of skills and the second by the demand for work among millions who may be on the verge of starvation. It tends to produce a buyers' labor market, one in which unions have little purchase on power. This in turn has grave consequences for the political parties based in part on working-class votes and organization.

It is within this changed framework that the continual process of melding divergent populations into a single society continues. The major policy dilemmas that arise derive from the broad conditions of change. First, shall the borders be closed to further immigration and, if so, how? Second, how shall a large and growing labor force, freed from the characteristic brute labor of previous civilizations, be used for humanly valuable work?

Clearly, the demand for immigration will not disappear while the pressure of population on sustenance approaches famine in today's huge societies; clearly ethnocentrism, persecution, and flight are not on the decline. From the Huguenots fleeing from France to the "boat people" of Vietnam, it is a continous and probably increasing aspect of civilization. And at all times there will be the thousands of aspiring youth who are trapped in stagnant societies where no future seems likely for them: Why would they not want to go to the center of the world?

Yet the United States cannot be the employer of last resort and the refuge of last resort for all these troubled populations without changes in its nature, its formulas for citizenship, and its basic values.

And with respect to the demand for labor, it is clear that much of the extraction and fabrication of mass consumed goods can be automated cheaply and effectively. The growth of the service area is not to be deplored, but its useful growth demands provision of the stunted public services, as John Kenneth Galbraith noted long ago. Until the "human use of human beings" is taken seriously as a commonsense axiom, we will have millions on the dole while the society's infrastructure collapses and its school systems revert to barbarism.

Thus the new ethnicity, one that has been brought about by new values on old variables, forces new thought upon policymakers faced with new problems. For the social scientists it also requires some rethinking of the basic aims of social science, as well as the basic nature of our subject matter—the behavior of the human species. We are recalled to the problematic nature of human societies as meaningful entities, and to the growth of civilizations as a synergetic process resulting from communication and cooperation among diverse peoples in urban centers. But such civilization is possible only with a moral order, and as assimilation itself is primary to a moral order, it must continue to be of central importance in our social theory. However, assimilation results from processes of acculturation, that is, two-way influence between the new immigrants and the older order. It is time to take this seriously and to measure what in the nature of American culture is in process of change at this moment, in Los Angeles, Texas, Miami, the vast conurbation of New York. What minimum order must survive for the functioning of a society that includes so many different groups with such various histories?

As an *envoi,* it is well to note that much clarifying and creative thinking was produced by the older study of ethnicity and, if we go back far enough (to Robert Park himself), the connections among migration, ethnicity, and nation building. The lesson seems to be, however, that you do not develop unchanging generalization about society through imagining an unchanging society, but rather by seeing, as Windelband noted long ago, "that which is unchanging in the nature of change."

University of Wisconsin—Milwaukee   *Scott Greer*

Introduction

LIONEL MALDONADO
JOAN MOORE

☐ OVER THE PAST few decades the nature and character of ethnicity in American cities has changed profoundly. This can be seen vividly in the shift in the origins of immigrants: Between 1960 and 1964 nearly two-thirds of the annual immigrants entered from Europe and Canada (45 percent and 12 percent, respectively). Only a decade later, between 1970 and 1974, this rate was cut in half: Fewer than one-third of the new arrivals came from European nations and Canada (28 percent and 3 percent, respectively). The flow from Asia and Latin America had increased dramatically; these became the majority of arrivals. These newest immigrants contributed an even greater cultural diversity to modern America than the people from Central, Southern, and Eastern European countries at the peak times of their arrival, between 1880 and 1920 (Bryce-Laporte, 1982). This altered state has important and far-reaching implications for the urban landscape of America.

The basic change is easily marked; it began with the new Immigration Act of 1965, a most comprehensive restructuring of U.S. immigration legislation. The new act replaced the highly biased and restrictive National Origins Quota Act of 1924 and tried to bring some order to the tangle of legislation grown up around the 1924 act. It repealed the historic favoritism for Northern and Western European nations, abolished all remnants of Asian exclusion, and further elaborated the preference system of the McCarren-Walter Act of 1952. Family unification was the main target, with secondary allowances for occupational skills and refugee status. Eastern Hemisphere nations were allowed 170,000 immigrants annually, with no nation allowed more than 20,000. (The elaborate system of preferences is shown in Table 1.) For the first time the Western Hemisphere fell under a ceiling of 120,000 persons, though without national limits. In 1978 both hemispheres were

TABLE 1   Preference System of the Immigration Act, 1965

First preference:	Unmarried sons and daughters of U.S. citizens.
	Not more than 20 percent.
Second preference:	Spouse and unmarried sons and daughters of an alien lawfully admitted for permanent residence.
	20 percent plus any not required for first preference.
Third preference:	Members of the professions and scientists and artists of exceptional ability.
	Not more than 10 percent.
Fourth preference:	Married sons and daughters of U.S. citizens.
	10 percent plus any not required for first three preferences.
Fifth preference:	Brothers and sisters of U.S. citizens.
	24 percent plus any not required for first four preferences.
Sixth preference:	Skilled and unskilled workers in occupations for which labor is in short supply in the United States.
	Not more than 10 percent.
Seventh preference:	Refugees to whom conditional entry or adjustment of status may be granted.
	Not more than 6 percent.
Nonpreference:	Any applicant not entitled to one of the preferences.
	Any number not required for preference applicants.

SOURCE: Keeley (1979: Table 2).

combined under a single preference system; the per nation limit remained and the total world ceiling was kept at 290,000.

But these changes did not end the national debate, particularly because of the American involvement in Southeast Asia and the persistent worry over communism. Pressing demands by refugees brought the Refugee Act of 1980, which increased the annual quotas for immigrants and refugees to 320,000 and 50,000, respectively.

This is the background for this book on the new immigrants. Our concern is not with immigration change or reform. We take as our assumption the massive reshaping of our cities by the millions of new arrivals and argue that neither American society nor its scholars has yet recognized this change or dealt with it. Nor are we concerned with the characteristics of the immigrants from particular nations—Mexican Americans, Koreans, or others. Rather, we must go beyond the case study approach and deal with the ideas and institutional structures that interact with and are affected by these newcomers. We are aware that nearly all social research in race and ethnic issues is dominated by an intense interest in Black-White relations. It is not necessarily true that the paradigms of such research will offer insight into the social and economic experiences of the newest arrivals.

The book is organized into two sections. In the first section, we will build a context for the study of contemporary racial and ethnic issues in American cities. To this end, Margo Conk describes the main currents in American thought and American political life that have generated changes in immigration law. America's early industrialization brought tremendous needs for labor; it was not possible for the need to be supplied by natural increase in the population. Hence there were few limitations on entry. But the children of each successive wave of immigrants began to feel possessive about their native land. And once the growth of heavy industry reached the point where a large labor force was no longer needed, nativism could, finally, generate the restrictive legislation of the 1920s.

Conk's essay suggests that an effective view of immigration is that of a flow of labor, moving from an economy where it cannot be absorbed to another where it is in demand. It also suggests how important the role of the modern state may be in ensuring the labor supply necessary for economic development. The state administers legislation that controls the utilization of extranational labor pools. The extent to which the domestic supply can meet the demand is always important. Thus international migrations are much more than merely the sum of individual motivations. Rather, migrations reflect global forces of an international stratification system with an asymmetrical relationship between states. Human capital produced in one nation is recruited to another.

Today, the character of the flow has changed. Morrison Wong profiles the newest arrivals. He observes that their greatest impact (after the Immigration Act of 1965) began in 1968, and he contrasts them with the older immigrants. The differences are obvious and important. First, there is a dramatic increase in the absolute numbers—a doubling in the twenty-year period from 1961 to 1981. Second, the rate of increase is accelerating. Third, there are major changes in the origins of the new immigrants. There is a very substantial decline in the numbers and proportion from European countries; a corresponding increase in numbers and proportions of Asians (particularly of ethnic groups quite new to the United States), and an increase in the numbers (but a decline in the relative proportion of the total) from Mexico and other Spanish-speaking nations.

There are important differences between the older immigrants and the new arrivals. The new immigrants are younger, tend to come from urban backgrounds, and tend to settle in urban areas. There are significant educational and occupational variations among the various groups—some below U.S. averages and others higher. The vast majority are of prime working age and find work. Thus American society clearly has borne no costs in educating and otherwise preparing these individuals for the work place. But the new arrivals bring their families, and this has implications for the schools and educational policy.

Tracing the same arrivals into the particular metropolitan area of Los Angeles is the task of the essay by Philip Garcia. Because it is a major settling point for the new urban immigrants, Los Angeles's experiences may be as useful for the study of the immigrant impact on American cities as were the experiences of Chicago in the 1920s. Garcia follows the racial and ethnic changes in Los Angeles neighborhoods since 1950, concentrating on the period since 1965. He observes the immigrants' pattern of initial settlement in older sections, then a gradual penetration into other nonimmigrant areas. The newest arrivals, along with Blacks, consistently remain more isolated from Anglos than did the European immigrants. He also notes that by contrast to the experience of European ethnics, socioeconomic status seems to have no effect in reducing segregation or stemming the exodus of Anglos. It may be that as America's racial and ethnic heterogeneity increases, so does its segregation. Unlike earlier times, the upward social movement of new arrivals apparently does not increase their acceptance by Anglos. The attitudinal data presented by Garcia seems to suggest a new phase in intergroup relations that is marked by deep and pervasive antagonisms that are less susceptible to factors of social class.

Jose Hernández draws on scattered research and intensive obser-vation in the Chicago and New York metropolitan areas to suggest a new typology of urban immigrant community building. But it is the data on new immigrants that mostly concern Hernández; he finds them to be dauntingly flawed. There have been two decades of massive efforts by demographers (including Hernández) to reform census data to reflect something other than the most archaic of nineteenth- and early twentieth-century racial bias and naiveté. The 1980 census made some major improvements, but many of these refinements will be eliminated in plans for the 1990 census. It will then be virtually impossible to do the analyses so badly needed to modify the older paradigms of ethnic ecology. Furthermore, Hernández argues, this new immigration includes so many undocumented individuals and so many other "anomalies" that the normal data collection techniques of other agencies are often invalid. Thus the new immigration has had a major impact in making the conventional sources of demographic information even less useful than before. Hernández believes this pattern is recognized by few demographers—and proposes that the new and old minorities who need accurate data must collect their own. To this end he proposes a number of procedures that must be followed, with particular attention to pitfalls in conventional approaches.

The second section of this book is concerned with the reciprocal impact of new immigrants and institutions. Interestingly, some of the new immigrants carry with them the personal and financial resources to flourish independently. High levels of education, good jobs, some financial resources, and strong networks both here and in the home nation are involved. Nonetheless, such persons face many strains in resettling in a very new environment. Even very sophisticated Asians and Africans, for example, must adjust to new culture patterns.

But most immigrants, now or in the past, are needy. For those without resources, American institutions are critical. It is noteworthy that all the essays in this second portion discuss the new immigrants in the context of the needs of native-born U.S. minority groups of non-European ancestry. This (as implied by both Garcia and Hernández) is not simply a matter of race and racism. The new immigrants, and especially the poorest among them, locate in the portions of cities that are most subject to "white flight." In such areas the existing institutions are the most pressured. Ecology, then, is an important factor, and nowhere is it more critical than with American schools. Ricardo Fernández and William Vélez document the devastating impact of city schools on both native-born minorities and newcomers. Tragically, the special programs developed during a more affluent period of the welfare state are eroding rapidly. New schemes for "second-tier" diplomas for young people who cannot pass minimum competency tests are among the innovations that seem likely to restrict opportunity and make a segmented larbor market a permanent feature.

But although the idea of the "segmented labor market" is useful, it is too simple a concept for the economic status of new immigrants. It is particularly important to understand this because some politicians claim that new immigrants take jobs away from old minorities, especially Blacks. Marcia Freedman argues that the new arrivals are not in direct competition and, further, that this misconception is the result of an excessively shallow notion of what is happening in the labor markets of American cities. Immigrant entrepreneurship is important. At one extreme, for example, the Cuban "enclave economy" in Miami is actually a self-contained subsystem. Elsewhere, ethnic restaurants give foothold jobs to immigrant relatives, and there are many tiny niches in odd corners of the economy. Immigrant-owned and marginal garment shops employ immigrant women. There is no extensive counterpart among Blacks. The fact that immigrants own enterprises using low-paid, unsheltered immigrant workers lends credence to the notion that labor supply creates its own demand. It is quite possible that without these sweated immigrants, jobs would locate outside the United States. But these patterns are also related to a general weakening of the labor market, in which such interstitial activities flourish. (Small business enterprises are very unstable.) The sheltered occupations are declining in importance, and the shelters in private industry are weakening.

This background suggests that native minorities and new arrivals are not really competing for the same kinds of jobs. Freedman believes that native minorities have advantages over immigrants in strong labor markets. Thus, public sector jobs and sheltered occupations with strong unions are important aspirations for native minorities, but these jobs are declining. Immigrants do well in more loosely structured markets. Although these realities add complexities to the idea of the segmented labor market, they must be addressed.

Eleanor Miller and Lynne Kleinman are interested in immigrants and American patterns of social control. They note that criminality has always been a preoccupation in immigration policy; historically, criminality was one of the earliest reasons for barring would-be immigrants. Criminality was associated with certain "racial" groups. And, second, at least one reason for the development of urban police forces during the nineteenth century was to control the disorder linked with immigrants. Further, Miller and Kleinman suggest that crime is beginning to be internationalized in a manner not true of earlier groups. Ethnic crime has always involved an element of work or business; now the illicit economy is operating on a scale that our agents of social control are not prepared to handle.

In his essay, Robert Magill reminds us that, as with the police, most welfare help is managed at the local level and thus depends upon local definitions of a situation. This is a legacy of a social welfare institution that was associated with aid to the earlier waves of immigration from Europe. The values underlying many welfare policies and programs are collectivist and find the source of problems in society itself. In this respect, Magill argues, social welfare runs at odds with the dominant value system of American culture, which emphasizes individual achievement and tends, therefore, to blame the needy for their troubles. Against this background, the new immigrants are very diverse in terms of their needs and in terms of their values. In general, the social welfare institution has just begun to notice the new people and to come to terms with their new problems. Tempering this response is the reappearance of old problems and budget cuts that seem much more urgent than the new problems of new populations.

William Liu and Elena Yu offer a definitive survey of the present status of mental health services and ethnicity, suggesting after an exhaustive review of existing research that mental health facilities are totally inadequate for the needs of new immigrants. This careful study concludes that, in particular, the mental health problems of Asians cannot be handled adequately by existing theoretical formulations in race and minority studies. Among the other suggestions in this sophisticated review is a warning that the inability of new immigrants to be identified in specific communities seriously impairs the usefulness of the mental health "catchment area" system. Most useful in the Liu and Yu essay, perhaps, is their general survey of the failures of current mental health delivery services to ethnics and the resulting warning that it will be even less effective for the new arrivals.

Then there is the question of political power and the new immigrants. Steven Erie offers an important revisionist view of the American urban political machine, noting that its usefulness will be extremely limited, and much less than is generally assumed. Observing that the dominant Irish machines of past and recent times were careful to mobilize new arrivals only in very selective fashion for tangible but limited rewards, Erie concludes that these are institutions designed for social control.

The new immigrants will have to cope with a political mechanism that is quite well equipped to control them. Even the benefits of a welfare state and greatly expanded social programs are unlikely to reach them without important political concessions to dominant political forces. This was largely true for Hispanics and Blacks and there is no immediate reason to expect that the "rainbow theory" in the form of urban political machines will share limited resources with the new immigrants.

It is interesting that the constant references in these essays to the early work of the Chicago sociologists remind us that the very foundations of urban sociology began with studies of immigrants. Immigrants and their children were the gang members studied by Thrasher (1963); Polish immigrants were studied by Thomas and Znaniecki (1918). The Jews of *The Ghetto* (1928) studied by Louis Wirth were immigrants, and Robert Ezra Park was directly concerned with the *Immigrant Press* (1922). Furthermore, this research then, as now, was done in a very politicized environment. Immigration and immigrants were major public issues of the period. Conk and Miller and Kleinman discuss the racist works of that time, reminding us that many political movements of the era were deeply xenophobic, and, in fact, sociologists sometimes were directly involved. From the University of Wisconsin, E. A. Ross espoused restrictionism as early as 1914 with *The Old World and the New*. Some Ross arguments prefigured modern restrictionism, claiming that unselective immigration would destroy the nationhood or spiritual unity of the American people and thereby destroy democracy. "Muddled mixing begets absolutist government," according to Ross (1922: 6).

The political climate also directly affected work of the Chicago empiricists. The famous study of immigrant adjustment by Park and Miller (*Old World Traits Transplanted*, 1969/1921) was funded by the Carnegie Corporation and originally was to have been a study of the loyalty of ethnic Americans in World War I. It was carefully retitled "Americanization studies" to avoid possible negative repercussions for immigrants. Later, however, Park was able to reject liberal pressure. In 1924 religious liberals in the New York-based Institute of Social and Religious Research appointed Park to study Asians on the West Coast. Park pursued a firmly disinterested line and reported in his "Tentative Findings" that tension had dropped since the Immigration Act of 1924 excluded all Oriental laborers. The sponsors wanted the conclusion deleted, but Park refused (Matthews, 1977: 113-114).

But none of the Chicago empiricists took a public stand on immigration reform. This may have been necessary because Chicago industrialists, among them Judge Gary, were violently opposed to restrictionism, although many felt it would reduce radicalism (Leuchtenberg, 1958: 205-206). This was important to the researchers because in 1922 they were negotiating with the Laura Spelman Rockefeller Foundation for funding. A portion of the funding was contingent on raising money in the community. Annually for about ten years Chicago civic and welfare agencies joined with the University of Chicago in raising $25,000. This sort of situation does not encourage public partisanship with its potential for alienating local donors.

The Chicago sociologists were also careful to divorce themselves from early development of the social work institution discussed in Magill's essay. Ernest W. Burgess summarized, "Chicago had been flooded with wave after wave of immigrants from Europe." And he noted the importance of public opinion: "By the time our studies began . . . public sentiment had crystallized into rather firm prejudice and discrimination against the new arrivals from Eastern and Southern Europe." He disavowed prejudice, as did Park: "The social scientists at the University of Chicago did not share, for the most part, the prejudices against these people that were commonly expressed." He acknowledged that the very earliest research was geared to combat these prejudices: "Much of the earliest 'social research' was little more than the discovery and reporting *to the public* that the feelings and sentiments of those living in the ethnic slums were, in reality, quite different from those imputed to them by the public." But then he cast off this earlier work and the ameliorist and activist tradition in favor of a new model of policy-neutral scientific research. "By the early 1920s this 'social work' orientation had given way . . . to an ambition to understand and interpret. . . . Although the objective was scientific, behind it lay a faith or hope that this scientific analysis would help dispel prejudice and injustice and ultimately would lead to an improvement in the lot of slum dwellers" (Burgess and Bogue, 1964: 4-5).

One after another, the essays in this volume directly confront the irrelevance of the Chicago models in describing the complexity of new immigrants and old minorities in American cities. This is most obvious with the ecological models. On the Burgess zonal hypothesis of immigrant settlement, Garcia finds some support for the old pattern. Hernández argues that community types are more diverse than can be managed by mere counting of proportions, as given in census data. Liu and Yu argue that although remnants of territorial segregation persist, communities are now defined by many other forces than simple residence. They note, furthermore, that the application of Chicago ecological models to mental health "catchment areas" is seriously out of touch with reality.

The Chicago models were accurate for their time. They were built on a large body of hard work in the field. However, the essays in this book make it clear that they have outlived their usefulness in contemporary American cities. The complexity of new immigrants and old minorities in late-twentieth-century cities cannot be comprehended without yet another large body of hard work in the field, nor can effective policies be developed until we understand what is happening.

Part I

Historical, Demographic, and Ecological Considerations

Immigration Reform and Immigration History: Why Simpson-Mazzoli Did Not Pass

MARGO CONK

☐ IN THE LAST DAYS of the 98th Congress, Republican Senator Alan Simpson of Wyoming and Democratic Representative Romano Mazzoli of Kentucky acknowledged the death of the Simpson-Mazzoli immigration reform bill. That bill had come within a hair's breadth of becoming law. One bill had passed the Senate overwhelmingly in May 1983. A very different version had passed the House after a week of intense debate in June 1984. A conference committee struggled through the summer and early fall to fashion a compromise that would satisfy the differing concerns of the House, the Senate, and the president. It was no easy task; commentators declared the bill all but dead many a time in those months. Nevertheless, Simpson and Mazzoli doggedly kept reviving it, working out compromises on each of the bill's many sections, working against the inevitable deadline of the adjournment.

The House and Senate versions differed in their treatment of "amnesty" for illegal aliens. The Senate bill would have granted amnesty to those aliens arriving before 1980. The House version dated its amnesty provision 1982. Both bills provided for sanctions for employers who hired illegal aliens. The House bill contained a controversial "guest worker" program termed "rent-a-slave" by opponents such as Representative Henry Gonzalez of Texas. The House bill also added civil rights protections to deter discrimination against foreign-looking job seekers. None of the compromises pleased all the parties to the debate, but Simpson and Mazzoli kept the conference going by reminding their colleagues of how long it had taken to get this far. Immigration reform, most commentators could agree, was an issue that had to be dealt with. It had taken years of research and study, and years of political compromise, to fashion the bills now under debate. Few politicians relished starting all over again with a new Congress. Get the matter behind them, reform advocates reasoned. Even a somewhat defective law would be better than the chaotic situation currently at hand.

In the end, these efforts failed and Simpson-Mazzoli died on a 15 to 13 vote in the Conference Committee. The vote rejected a Reagan administration demand that federal aid to local governments for new services deriving from the amnesty provisions not exceed $1 billion a year. Immediately a flood of recrimination began in the press as the various sides in the debate blamed one another for what had happened.

For quite some time after that fateful vote, interpretative pieces continued to appear. It was Reagan's fault for refusing to pay for the amnesty provisions. It was Mazzoli's fault for not working effectively with Congressman Peter Rodino of Newark. It was the Democratic leadership's fault for playing politics and obstructing the bill through the election year. It was everyone's fault for not rising above special interests to solve this difficult national problem. All true.

But the fascinating thing about the debates over Simpson-Mazzoli was the sheer cacophony of opinions and positions. Besides the small group of politicians—such as Simpson and Mazzoli—who have made immigration reform their specialty, it was hard to find a coherent set of advocates or lobbyists on one side or another. Supporters included groups specifically organized around immigration issues such as FAIR (the Federation for American Immigration Reform). This group sued the Census Bureau in 1979 to prevent the illegal alien population from being included in the 1980 reapportionment. Many members of Congress simply felt that something had to be done about the defects in current law that allowed but did not acknowledge the massive immigration along the southern border. Such support was broad but shallow. Opponents were an equally diffuse grouping. They included the AFL-CIO, the American Civil Liberties Union, the U.S. Chamber of Commerce, and Hispanic organizations.

It is this sheer inability to find much of a coherent policy thrust on the issue that is, I would suggest, the key to understanding some of the deeper reasons that Simpson-Mazzoli failed, and why in general Americans find it so difficult to fashion effective immigration policy. The bottom line is that Simpson-Mazzoli was about immigration restriction, and most Americans do not particularly want immigration restriction. Despite all the talk about "controlling the borders," Americans do not find the presence of 6 million or more undocumented or "illegal immigrants" sufficiently dangerous or problematic to do anything about it. Illegal aliens are a nuisance to census takers; they are "problems" to local government officials, who can not gauge the need for local services. But they are also willing and cheap workers, politically quiescent, and, for most parts of the nation, a very distant problem.

The purpose of this essay is to explore the long-term patterns in the American economy and social structure that encourage immigration and make restriction extremely difficult. Since the current wave of immigrants is at least the "fourth wave" (Muller, 1984) according to a recent influential study of the effect of the current immigration on Southern California, the history of previous immigrant waves and previous efforts at restriction provide the means to evaluate these patterns. They indicate that restriction has occurred only during periods of severe societal crisis. We are clearly not there at the moment, but perhaps, as we explore the past, we will be able to conceive of the conditions that might encourage the passage of another Simpson-Mazzoli bill.

THE CONTEXT

Throughout most of the nation's history, Americans considered immigration an unmitigated good. We like to consider ourselves "a nation of immigrants"—the descendants of the generations of immigrants from the Old World who were going to find the good life in the New. This sentiment corresponded quite well with the generally accepted principles of American social and economic development. People create wealth; hence more people create more wealth. Further, the United States has historically been a labor-short nation; despite extremely high rates of fertility and relatively low mortality, the country could not satisfy its demand for labor through natural increase. Foreign immigration was the only other source. Until the twentieth century, federal and state governments not only did not restrict foreign immigration, they followed policies that encouraged it. Only in the 1920s did the United States decisively abandon its traditional policy of unlimited open immigration and begin to pass restriction laws that regulated the flow in one form or another.

These favorable attitudes and the resultant government policies led to a truly enormous flow of humanity into the United States. Between 1820 and 1980, over 49 million immigrants came to the United States according to the official statistics. During that time, the U.S. population grew from under 10 million to 226 million.

Further, Americans have considered the right to move and settle anywhere in the country one of their most important individual freedoms. For almost the first two centuries of European settlement, communities remained close to vital water transportation routes along the East Coast. In the nineteenth century, though, Americans poured over the Appalachians, established farms on the prairies and plains of the Midwest and Far West, and built railroads connecting settlements on the Pacific Coast with those in the East. They also built great cities out of empty stretches of plain and prairie, drained swamps to build harbors, created great industrial centers in the course of a generation. To accomplish such feats of development, countless sons and daughters of Easterners moved hundreds of miles west to settle new farm lands, or migrated into nearby or distant cities to build an urban world (Nugent, 1981). Chicago is perhaps the extreme case of this phenomenon. Its population was about 30,000 in 1850; over a million and the second city in the nation by 1890. The census of 1890 reported that 450,666 Chicagoans had been born abroad.

In short, White native Americans were also migrants—to the west, to cities, to new environments in which they met and formed communities with other migrants. The general volatility of the native-born White population paralleled the high rate of foreign immigration. The two factors have formed two of the underlying characteristics of American population dynamics throughout our history. Any movements to restrict immigration have thus had to challenge these basically favorable attitudes toward both internal migration and foreign immigration. They have had to answer the question as to why more people would not be beneficial to the society and more precisely, why sparsely populated and newly settled areas of the country should not have the benefit of migrants as have older regions. The restrictionists have historically been on the defensive, arguing against the general American predilection to favor a liberal immigration policy.

The period of European settlement of the geographic area that is now the United States spans almost four centuries, from the early modern period to postindustrial society. For most of those four centuries, agriculture was the dominant economic activity. Rural society was the dominant social form. About 100 to 150 years ago, the United States began to industrialize. By 1880, the Northeast had become an urban industrializing region; by 1920, the Census Bureau classified half the U.S. population as urban.

Large-scale foreign immigration characterized both the agricultural and the industrial periods of American history. During the preindustrial period, the numbers of immigrants were lower in absolute terms, but their relative impact on the society was perhaps greater. During the last 130 years, immigrants primarily fueled the industrial sectors of the American economy as wage laborers. There they had a major impact on the shape of the society, though numerically they formed a relatively smaller portion of the overall American population.

In terms of the long-term development of American society, seventeenth-century Americans made crucial decisions about the "structuration" (Giddens, 1973) of class and race relations in the infant colonies that have determined the character of social relations ever since. As is generally known, two radically different patterns developed. In the North, radical English Protestant sects made "errands into the wilderness" (Miller, 1956) to create model communities designed to preserve the theological purity that was threatened by the Anglican church. The New England colonies they founded were primarily agricultural societies, but a combination of factors—including poor land and the need to search for other economic activities to sustain the infant colonies—led to a diversification of their economies. The tight theological communities evolved into significant urban settlements. By the end of the seventeenth century the prosperity of these colonies rested upon their mercantile activities. The immigrants who continued to fuel these colonies found work not only on the land, but in the artisanal trades or shipping activities of the coastal cities (Bailyn, 1955; Katz and Murrin, 1983).

In the South a totally different pattern developed (Morgan, 1975). The southern colonies were founded as joint-stock companies designed to bring profits to their investors. They were unsuccessful until the 1620s, when the Virginians discovered the cash crop, tobacco, and were able to put the colony on a self-sustaining basis. The locally grown tobacco fetched a relatively high price in European markets. Men who were willing to migrate to Virginia to grow tobacco could get rich. Still, there were problems with the infant plantation economy. After the initial tobacco boom of the 1620s, the price fell, and it took much longer to make big money as a tobacco planter. Virginia never did seem capable of attracting a class of stable, God-fearing immigrants of the New England variety. It was a decidedly unhealthy place for Englishmen in the seventeenth century; the death rate during "seasoning" was shockingly high (50 percent). And it was well-known for its coarse frontier lifestyle, a characteristic that made it the forerunner of many American boom economies that were to follow it over the centuries.

Virginia, in short, developed a chronic "labor problem." Until the late seventeenth century, the Virginia planters relied chiefly on white English indentured servants, but this system had inherent problems. Because Virginia was such an inhospitable place at the time, servants expected the "headright" in land that they were to receive after seven years labor to allow them to become planters themselves. Over time, therefore, the indentured servants became the competition to their former masters. Meanwhile, the price of tobacco kept falling; the older planters used their control of government offices, the best land, and the legal system to maintain their economic and political hegemony over the colony. The newer up-country planters began to chafe at the exploitation of the elites.

Things came to a head in the late 1670s and early 1680s when the colony was rocked by civil war. The Americans faced their first "immigration problem," because the Virginians had built an economy that bred social instability. The planters had yet to find a means to satisfy their continuing needs for new labor while peacefully integrating those new immigrants into the social structure. After Bacon's rebellion, the planters made a decisive shift away from the importation of white English indentured servants as laborers. Instead, they turned to Black slave labor.

Black slave labor had been introduced in small numbers into the tobacco economy in the mid-seventeenth century, and White laborers and Black slaves had worked side by side in the tobacco fields. Blacks constituted about 10 percent of the Virginia population in the last quarter of the seventeenth century—that is, between 3,000 and 6,000 people. Despite what might seem to be the obvious advantage to slave labor—namely, that slaves worked for life—the planters did not immediately choose slaves. Slaves cost more than English servants; hence, given the heavy mortality rates during seasoning, they might not be worth the investment. Only after the rebellions, when the drawbacks of English labor were all too evident and life in Virginia was a bit healthier, did the planters reevaluate their attitudes toward the two forms of labor and choose to invest in slaves. As they did so, they created the racial dividing line that has characterized social life in the United States ever since.

Between 1700 and 1750, Virginia imported 45,000 slaves and her slave population grew to 100,000. Blacks constituted 44 percent of Virginia's population in 1750; on the eve of the Revolution, she was the most populous and wealthiest of the original thirteen colonies. Virginia had found a solution to her labor problem. The widespread forced migration of African slaves provided a cheap, reliable source of agricultural labor. The system perpetuated itself, because slaves served for life and the children of slaves were slaves.

Thus, from the point of view of immigration, two trends developed during the colonial period. First, English immigrants came in search of a fortune, to practice their religious beliefs unharrassed, to fulfill a sense of adventure, or simply to improve their economic well-being. Given the abundance of land and the danger and hardship of colonial life in the seventeenth and early eighteenth century, the authorities, whether the Crown or the local elites, found it difficult to exploit these immigrants too severely. Overly harsh treatment of servants could limit the flow of immigrants or precipitate rebellion. By the eighteenth century, White European immigrants were generally guaranteed individual liberty and a reasonable potential to acquire land or property either on arrival or after the completion of an indenture—that is, to become "free-born Englishmen."

Second, Black slaves had no such possibilities. They labored for life, had none of these rights, and bore children who were also slaves. Only rarely were slaves emancipated by their owners or allowed to buy their own or their relatives' freedom. The result of such a stark system was that there was almost no intermediate group of ex-slaves, or free Blacks, who might provide a bridge between slavery and freedom. Over time Whites and Blacks continued to migrate and be born into this social and economic system and to assume the roles defined by the established patterns.

In the eighteenth century before the American Revolution, somewhere between 700,000 and 900,000 immigrants (including slaves) came to the thirteen colonies. During that period the population grew from 250,000 to over 2 million. Despite these high numbers, Walter Nugent (1981: 49) suggests that only a quarter of the population growth resulted from immigration. The rest was natural increase.

The Revolutionary Era altered the relationships among slave and free, and among the various social orders of colonial American society, in significant ways. Most critically, the Revolution led to the gradual abolition of slavery in the Northern states, and the first major debate about the moral, legal, and political disadvantages of slavery as a labor system. Southerners defended the system, but the new national government made some significant efforts to limit its expansion. The international slave trade was prohibited after 1808. Slavery was outlawed in the Northwest territories. Slaves were counted as three-fifths of a free person for the purpose of apportioning representation in the House. And there was some indication that the tobacco economy, which had provided the economic rationale for slavery, was in decline. As the infant United States entered the nineteenth century, intelligent statesmen could hope that the system of racial slavery might die gradually (Davis, 1975).

White laborers also gained additional rights as a result of the political upheaval of the Revolutionary Era. Indentured servitude slowly died out in the late eighteenth and early nineteenth centuries. It was replaced by apprenticeships and free labor contracts. Americans sought to become yeoman farmers or property-owning mechanics, both forms of "free labor" (Foner, 1970). Wage labor was an appropriate status for young men who, contemporaries felt, could aspire to become property-owning farmers or mechanics as they got older. During the first third of the nineteenth century, property qualifications were dropped for voting and office-holding. The United States became the first nation of the world to institute universal White male suffrage.

During the first half century of the American republic, the U.S. population grew phenomenally, from 3.9 million according to the first census in 1790, to 17 million in 1840. This growth was almost totally the result of natural increase, as immigration was negligible at the time. By way of comparison, the birthrate in the United States in the early nineteenth century was, according to Walter Nugent (1981: 55), "higher than that of any country in the world in the 1970s." Americans interpreted this population growth optimistically—as evidence of the virtue and vitality of their new "republican" form of government. Population spilled across the Appalachian Mountains; thirteen new states were settled in the west between 1790 and 1840. By 1860, over 40 percent of the American population lived in the new states.

Historians have noted that the western states were generally settled by migrants from eastern states of roughly the same latitudes. New Englanders, for example, migrated to western New England and New York in the first third of the nineteenth century; they moved to Michigan and Wisconsin in the 1840s and 1850s. Similarly, southerners settled Mississippi, Alabama, and Louisiana. States such as Indiana, Ohio, and Illinois received migratory streams from both North and South. And, most crucially, the migrants brought with them the institutions and social forms from the East. Southerners carried the slave economy to the West; northerners brought the free labor system.

Both systems prospered during the first half of the nineteenth century. In particular, the slave economy gained a new lease on life as planters shifted to cotton production after 1800. The efforts during the Revolutionary War era to discourage the expansion of slavery in the new republic proved ineffective in the face of the extraordinary growth of the Cotton Kingdom in antebellum America. Despite the ban on the importation of slaves after 1808, American slave owners found they could supply the new plantations in the West. Unlike the situation in all other New World slave societies, the American slave population reproduced itself at a strikingly good rate; there were enough slaves to work the older plantations in the East and settle the new lands of the West (Fogel and Engermann, 1974).

The effect of the early efforts to restrict slavery thus did not limit its growth. They merely channeled it into the newly formed southern states, and reproduced the settlement patterns evident in the East. By the 1840s, as in the earlier period, the United States displayed two vital yet radically different social systems. For a variety of reasons, including the development of a radical new abolitionist movement in the 1830s, Northerners and Southerners began to view one another with increasing suspicion during the 1840s, and question whether the United States could continue to develop with such different social and economic systems.

During that decade, two new elements were added to the social and political agenda that exacerbated these tensions. The first were the debates about territorial expansion and the "manifest destiny" of Americans to people the entire North American continent. The second was the return of significant foreign immigration to the United States. Both forced Americans to articulate their ideas about the shape of the future American social structure. And both ultimately caused Americans to begin to qualify and limit their faith in their ability to create a free, peaceful and abundant nation and to assimilate non-"Anglo-Saxon" races into the American dream. It is perhaps one of the deepest ironies of American history that a fully developed theory proclaiming the superiority of White Americans over other non-White and non-Anglo-Saxon peoples developed at the same time that democracy was expanding in the United States (Fredrickson, 1971; Jordan, 1968).

During the 1840s, United States territory expanded for the first time since the Louisiana Purchase of 1803. After negotiation with Britain, the sparsely settled Oregon territory was added in the Northwest. After the victory over Mexico in the Mexican War, Americans had to decide how much of Mexico's conquered territory should be annexed to the United States. For the first time, Americans decided against making major territorial acquisitions because of the racial and ethnic character of the indigenous population (Horsman, 1981). Mexico had a population of 7 to 8 million at the time, yet Americans took only the sparsely settled sections—Texas, California, and the portions of Mexico north of the Rio Grande, making 100,000 Mexicans citizens of the United States. As the *Louisville Democrat* put it at the time, Americans wanted "all the territory of value that we can get without taking the people" (quoted in Horsman, 1981: 245).

These territorial acquisitions also led to furious debates in Congress about the extension of slavery into the new states. Northern opinion began to crystallize around a goal of limiting the expansion of the slave economy. Southerners of course maintained that slavery was a matter of internal state policy and pressed for extension. These controversies almost paralyzed the national government in the 1850s, and did lead to the collapse of national political party system, the realignment of the Democrats as the Southern party, and the rise of the Republic party in the North and West.

At the same time, the Industrial Revolution, the growth of big cities, and foreign immigration began to change the way of life in the North. In the 1840s and 1850s 4.3 million immigrants came to the United States, or slightly more people than the 1860 slave population of 4 million. Of the foreign-born population in 1860, 70 percent were from Ireland and Germany. The heavy migrations were the result of abrupt political and social upheavals in Europe—the Irish potato famine and the failure of the German Revolution of 1848. The census also revealed that the immigrants went to some parts of the country and not others; 87 percent of the foreign-born lived in the free states in 1860. They also showed a strong proclivity to concentrate in cities. In New York and Pennsylvania, foreigners made up 44 and 31 percent of the population in large cities in 1860, but only 16 and 10 percent of the population in the remainder of the state (Abbott, 1926: 336). In the West, more than half the population in cities such as Chicago, Detroit, Milwaukee, and St. Louis were foreign-born, while in the rural areas of these states more than two-thirds were native. Immigrants thus tended to concentrate in manufacturing jobs, as railroad workers, or as common laborers in the cities. Though their presence as farmers was not uncommon, it was immediately obvious to contemporary observers that the greatest practical opportunities for work and a new life for most immigrants were in the burgeoning cities.

There were some politicians and social commentators who saw a threat to American institutions in these demographic patterns. They worried that the foreign-born could come to dominate the political life of the nation's large cities; they were concerned that the immigrants would take jobs away from natives. They worried that the Irish were Roman Catholic and thus hostile to the dominant Protestant culture. Germans seemed to wish to maintain their language and culture in the United States. Both the Irish and Germans drank whiskey, beer, or other alcoholic beverages to what many middle-class Americans thought was an excess. The immigrants had little sympathy with the temperance movements sweeping American society at the time. The result of these fears and concerns was the growth of a nativist movement in the 1850s. It coalesced around the Know-Nothing Party and led to an effort to stiffen naturalization laws, exclude Catholics from public office, and assert a virulent form of American nationalism (Billington, 1938; Higham, 1973; Abbott, 1926).

As the decade wore on, however, it became clear that the "impending crisis of the South" and the threat of a slave labor economy to the "free labor" system (Foner, 1970) would absorb all the political energy in the nation. The question of the future of slavery came to a head during the presidential election of 1860, as the slave states refused to accept the election of a Republican president. Over the winter of 1860-1861, eleven southern states seceded from the union. By spring, when Lincoln asserted federal power in the South, civil war became inevitable (Potter, 1976).

Ostensibly Lincoln went to war to "preserve the union," that is, to assert the power of the national government over the states. But below the surface, as the abolitionists continually pointed out, was the question of ending Black slavery and reconstituting the southern economy on a free-labor basis. Lincoln, along with much moderate Northern opinion, initially hoped to win the war without taking a decisive stand on abolition. But after a year and a half of brutal war, Northerners admitted they could not do so. Bowing to the pressure of the Radicals, Lincoln emancipated the slaves in the Confederacy as of January 1863 (McPherson, 1982).

The war unsettled social relations in the South, and by the middle of the war, Congress was debating just how to "reconstruct" the union and what the role of the "freedman" was to be in postbellum society. As the debates began during the early years of the war, most White Americans could not conceive of giving full political and civil rights to the freed slaves. By the end of the decade, events had forced Northern White

Americans to endorse not only an end to the slave system, but also citizenship rights, civil rights, and suffrage of Black Americans. They did so reluctantly, in an environment of racism that made many legislators and social commentators doubtful about the ability of Blacks to make use of their newly legislated rights, but they did lay the groundwork for the ideal of racial equality. And in the short period of the late 1860s and 1870s, when federal troops provided a modicum of protection to Blacks who tried to assert those rights, the southern states experimented with reconstructing the shattered cotton kingdom on a basis other than a slave labor system.

We know now that those experiments failed, that the Northern resolve to eradicate the power of the slaveholders waned, and that Congress rejected the more radical proposals for land confiscation and redistribution—"forty acres and a mule"—that might have enabled Black Americans to secure an economic base independent of their former owners. By and large economic power remained in the hands of the White elites who had led the move to secede. The South instituted a system of sharecropping or tenant farming in the late nineteenth and twentieth centuries that was designed to keep the agricultural labor force impoverished and tied to the land. The region never recovered from the devastation of the war; it fell far behind the rest of the country in wealth and other measures of social progress (Gillette, 1979).

The failure of Reconstruction further entrenched an ideology of racism in American social and political thought. Since many of the deeper goals of the Civil War were not realized, many Northern intellectuals and social commentators retreated even more from the egalitarian and radical republican views on human nature that had been accepted as givens in the prewar era. Further, the war unleashed one of the cruder phases of capitalist development in the North and West. The late 1860s and 1870s were filled with stories of political corruption, economic greed, and lowered moral standards. The infant labor movement, which had close ties with both the urban immigrant communities and the Republican party, began to assert its own version of equal rights—and claim that the burgeoning capitalist elites were corrupting the country by grinding the true "producing" classes. In this atmosphere, the coalition of abolitionists, businessmen, mechanics, artisans, and western farmers that built the Republic party began to fracture. As it did, the various elements of the coalition looked for someone to blame. Key sections of Northern intellectual opinion became increasingly conservative about the ability of the United States to assimilate both urban immigrants and Blacks (Keller, 1977; Montgomery, 1967).

Blacks were the first to feel the effects of abandoned ideals and racial prejudice. Northerners became increasingly pessimistic about the possibility of integrating Black Americans into the American mainstream. The hopes of the immediate postwar years were replaced by grimmer statements about the potential for social uplift and advancement (Fredrickson, 1971). By 1890, many social commentators had closed the books, so to speak, on Black Americans and had consigned them to a permanent second-class status in the United States. Richmond Mayo-Smith (1890) expressed a fairly typical view in his *Emigration and Immigration*. In his discussion of the demographic elements of the American population, Mayo-Smith defined three basic groups: "the descendants of the original colonists," "the immigrants since 1790 and their descendants," and "the negroes." "The negroes," he blandly continued, "are by birth and race and previous condition of servitude incapable of representing the full American capacity for political and social life." As "a legacy of the slave periods," they "will always be a problem for us." The only saving grace that Mayo-Smith could find in this unfortunate "national problem" was that Blacks had "displayed a docility and good nature since their emancipation which . . . made them a comparatively harmless, if not progressive and desirable, element in our national life" (Mayo-Smith, 1890: 64-65).

Immigrants were not initially subject to such derogatory evaluations of their role in their adopted country. Immigration resumed after the war was over, and business interests and governments did their best to lure immigrants to their areas. In the 1870s 2.8 million immigrants came to the United States; 5.2 million came in the 1880s. The immigrants continued to fuel economic development in the big cities and in the underpopulated West. In the latter half of the 1880s, though, crucial elements of American public opinion began to apply the rhetoric of racism to immigrants. They saw open immigration as a new "national problem" that could undermine the vitality and viability of American social, economic, and political institutions. There were several causes of this new concern (Higham, 1973).

The most important precipitating cause of the new pessimism about immigrants was the wave of strikes and labor agitation that swept across the nation in the late 1870s and 1880s. In 1877, for the first time, the United States was paralyzed by a nationwide railroad strike. The Knights of Labor and the American Federation of Labor both organized and assumed national importance in the early 1880s. Strikes broke out with increasing frequency in the early 1880s; the eight-hour movement culminated in a nationwide call for a general strike on May 1, 1886. In cities and towns across America, struggles between "labor" and "capital" seemed to be dissolving an old harmony between social classes. Anarchists and socialists called for the end to capitalism. The United States, which had long prided itself on its absence of sharp class antagonisms as in Europe, seemed to be facing a new and much less egalitarian social order (Montgomery, 1979; Fink, 1983).

Commentators of all stripes could see that the conflicts were worst in the big cities and that immigrants were concentrated in the cities and were active in the new labor organizations. In 1880, for example, half the working population in the nine big cities between New York and Philadelphia was foreign-born. In 1890, over half the working populations of Chicago, Detroit, Milwaukee, and Cleveland were foreign-born. Even the more conservative leaders of the labor movement suggested that unrestricted immigration, in the form of contract labor, was undermining wage rates for American workers and causing discontent and division within the labor movement.

In this atmosphere the ideas of the Know-Nothing nativist movement of the 1850s again began to appear credible. In particular, in a reversal of the Darwinian notion of the "survival of the fittest," old stock New England Brahmins worried that the native "stock" would be undermined by competition with immigrants. Francis Amasa Walker, the Gilded Age census superintendent and intellectual who popularized these ideas, theorized that the declining birth rates he discovered in New England were "caused" by immigration. Immigrants accepted a lower standard of living for their families. Americans, on the other hand, who wished to maintain a decent American standard, Walker theorized, limited the number of children they had (Solomon, 1956).

Finally, from the West Coast came the first concrete example of successful immigration restriction. After strong agitation from westerners, Congress passed the Chinese Exclusion Act in 1882. Chinese immigration was cut off for a period of ten years because of the supposed unfitness of the immigrants themselves. Chinese immigration to California dated from the Gold Rush days and had been controversial for many years. Many Chinese had been brought as contract laborers and worked under conditions that resembled slavery. Racial enmity toward the Chinese in California had led to a variety of discriminatory laws against laundries and other enterprises run by the Chinese. As Congress acknowledged the racial conflict on the West Coast, they established the principle that a people of a particular race or nation could be targeted for exclusion (Garis, 1927); with the Gentlemen's Agreement of 1907 the government extended it to the Japanese.

Thus by the mid-1880s, the pieces of policy and theory that would lead in later years to a full-brown restriction movement were in the air. The labor agitation of the 1880s called into question the conventional wisdom that immigrants were an unalloyed economic good. Walker's racial theories suggested that too much immigration was merely serving to lower the American birthrate and hence population growth rate. Immigration did not necessarily lead therefore to a net gain. And finally, the debates about Chinese exclusion led policymakers to see potential restriction policies in racial terms.

In the 1890s, these threads came together in the first serious immigration restriction movement in the nation's history. The Immigration Restriction League was formed in Boston in 1894 to agitate and lobby for legislation in light of these views. It set out to publicize the dangers of unrestricted foreign immigration and to develop specific legislative proposals to regulate the flow of humanity into the United States.

As later history indicates, the League was facing an uphill battle. They worked for over thirty years before they saw their theories become national policy in the National Origins Act of 1924. In that time 18 million more immigrants had come to the United States. The League and its allies faced formidable opponents who wished to maintain the traditional American policy of open immigration. Among the most important were the lobbies of the immigrant communities themselves, the business organizations representing employers of large numbers of immigrants, and the civic leaders who wished to lure immigrants to their local areas. Before the restriction movement could be successful, it had to win over, silence, or discredit this opposition.

This would prove to be both a difficult and protracted struggle and would require different approaches for the various sectors of the pro-immigration forces. The solidarity among the representatives of the immigrant communities had to be broken, and local leaders had to be made to see the "dangers" of unrestricted immigration. But perhaps the most powerful and most important sector of opinion to convince about the "evils" of unrestricted immigration was the business community. Employers in many industries had become accustomed to relying on a large supply of immigrant labor and had built their businesses around them. It would not be easy to convince the American business community that unrestricted immigration did not contribute to the overall prosperity of the nation. Businessmen knew that in the final analysis the immigration question was an economic one; immigrants were fundamentally workers—if not necessarily highly skilled, at least willing and malleable ones. Business organizations were unenthusiastic about immigration restriction because they understood that the American economy was built upon westward expansion and the development of railroads, cities, farms, and factories across the nation. An expanding population of workers kept the process going.

In particular, over the years the United States had built an economy that was geared toward the use of large numbers of unskilled laborers. This was partially the result of the peculiar pattern of American economic growth—that is, the exploitation of an entire continent for agriculture and industry that had hitherto been settled primarily by Native American populations. Forests were cut down. Land was converted to farms. Natural resources were abruptly and radically extracted. Such exploitation of nature utilized vast amounts of unskilled labor to clear the land, lay the railroads, build roads, cities, factories.

Further, the United States had historically been in the position of being an underdeveloped nation in relation to Europe. It had suffered from severe shortages of skilled workers and relied upon European imports for many manufactured products. By the nineteenth century, Americans no longer had to burn down their old houses to salvage the nails for their new ones, as they had done in seventeenth-century Virginia, but they still looked to Europe for first-class machinery, cutlery, china, textiles, books, and furniture. Americans had even devised their own solution to the problems of their backward manu- facturing sector. The "American system of manufacture"—initially designed for gun manufacture—used standardized and interchangeable parts to facilitate production. It became an accepted tradition of American manufacturing to simplify the product and the production process to accommodate to the character of the American labor force: a relatively short supply of skilled labor and a relatively good supply of unskilled labor (Morgan, 1975; Smith, 1977).

In the late nineteenth century, another twist was added to this principle of adjusting the manufacturing process and product to fit the character of the labor force. The industrial economy grew phenomenally in the years after the Civil War. The changes initially required building more and bigger factories, employing more workers, and thus producing more products. After 1880, as Daniel Nelson (1975: 4-5ff.) has shown, increases in scale necessitated changes in organization and technology because the larger enterprises simply would not function efficiently as autonomous "congeries of craftsmen's shops." Technological and organizational change in turn challenged the prerogatives of skilled workers—both in terms of their control over the day-to-day production process and in terms of the security of their jobs. As the work of historians such as David Montgomery (1979), David Brody (1969), and P. K. Edwards (1981: 92-93) have shown, the Gilded Age strike wave can also be seen as the response of skilled workers—most of whom were native Americans or Northern European immigrants—to regain their traditional control over the work process and to frustrate managerial efforts to restructure industry according to their vision of "efficiency."

Employers responded, somewhat predictably, by becoming even more convinced of the need to control what went on inside their factories. Older traditions, business journals suggested, such as sub-contracting work inside a factory, or tolerating the tramping artisan, had to be stopped. Managers had to design the production process, determine the pace of work, determine the appropriate level of production, and so on.

These ideas are best associated with Frederick Taylor, the founder of "scientific management." Taylor went to work in the 1880s in the Philadelphia metal-working industry (Midvale Steel) and discovered that an enormous amount of workers' time was, in his view, "wasted." Employers, he suggested, could get much more out of their workers if they analyzed the production processes, broke down complex tasks into their separate parts, and timed and set precise production standards for all jobs. Taylor's ideas spread slowly through American industry from the 1890s through the 1920s. They were perhaps best used and extended by Henry Ford after 1910 in his automated auto assembly lines. These in turn became the ultimate symbol of twentieth-century factory work (Nelson, 1980; Meyer, 1981; Gordon, Edwards, and Reich, 1982).

The effect of all this on the immigration debates was to reinforce the demand for unskilled immigrant workers. Unskilled immigrant workers did not have craft traditions to protect or ingrained notions of their accustomed privileges or rights. And they were cheaper to employ (Shergold, 1982). If unskilled non-English speaking immigrants could not be easily trained in the use of elaborate machinery, then the machinery and products would have to be adapted to the immigrants. Twentieth-century American mass production techniques and scientific management, "the most powerful as well as the most lasting contribution America has made to Western thought since the Federalist papers" (Drucker, 1954: 280), made unrestricted immigration the key to phenomenal industrial growth in the United States between 1890 and

1920. By 1907-1908, the majority of the male employees in metal manufactures, textiles, clothing, leather, oil refining, slaughtering and meatpacking, and sugar refining were foreign-born. (Nelson, 1975: 80). Clearly, manufacturers had little reason to kill the goose that laid the golden egg and restrict immigration.

In response, the voices of the restrictionists became shriller as the waves of immigrants kept coming. In the early twentieth century, Congress showed little interest in stemming the tide. Until 1916, even simple literacy test restriction bills failed in Congress or were vetoed by the presidents. Congress established the Bureau of Immigration in the Treasury Department (1891) to keep records on immigration, and set up depots like Ellis Island to process the arriving immigrants, but it was unwilling to go further to regulate the flow. Yankee intellectuals warned darkly that the nation could not successfully absorb millions of new immigrants, but the prosperity of the industries in which they worked and the cities in which they lived belied those warnings.

Nevertheless, the restrictionists were not totally unsuccessful in those years of heavy immigration in the beginning of the century. They did manage to build an elaborate theory of why immigration had to be stopped. The immigrants, they suggested, though economically valuable, were politically dangerous because they lacked the capacities to become equal citizens of their adopted nation. In particular, the restrictionists managed to borrow ideas from theories "proving" the inferiority of Blacks and extended their application to immigrants. The most powerful arguments justifying inequality in the United States were those devised to explain the discrimination aganist Blacks, Indians, and other non-White peoples. If immigrants could be considered as such a group, then it would be much easier to justify restriction.

Proving that the White European immigrants were racially different and hence deficient in the qualities necessary for American citizenship was no easy task, because, on the face of it, the immigrants were obviously White. Social theorists were not deterred by mere physical appearance and struggled to find some basis for separation. They found it when they noticed that the sources of the largest immigrant streams were changing in the late 1880s and that Northern and Western Europe (primarily Britain, Germany, Ireland, and the Scandinavian nations) no longer were sending the most immigrants. The "new" immigrants were preindustrial peasants from the nations of Southern and Eastern Europe. They spoke different languages, were predominantly Catholic or Jewish, and came from nations with "despotic" political systems. Such immigrants were indeed "strangers" to American institutions.

They looked different, and they seemed horribly poor and degraded as they arrived in the nation's cities looking for work. These "beaten men from beaten races," as Census Superintendent Francis Walker put it, would not have come to the United States in earlier decades when life was ruder and harder. This was not a natural migration of people looking to improve their lot. Only the unscrupulous advertisements and cheap fares of the steamship companies enticed these people to come to the United States. In the early years of the century, the infant eugenics movement drew upon these ideas and generated a whole spate of "scientific" studies of the racial character of the American population (Solomon, 1956; Higham, 1973; Conk, 1984).

After the turn of the century, the notion that these new immigrants were racially different from and inferior to older immigrants and the native stock began to seep into public consciousness. The New York State Factory Investigating Commission reported in 1912, for example, that in upstate factories, native Americans were referred to as the "white help." The foreign-born were "a separate class" (quoted in Nelson, 1975: 196). John R. Commons, the dean of American labor economists, wrote in 1907 that "the race problem in the South is only one extreme of the same problem in the great cities of the North, where popular government, as our forefathers conceived it, has been displaced by one-man power, and where a profound distrust of democracy is taking hold upon the educated and property-holding classes who fashion public opinion" (Common, 1907: 4). And in 1911, the prestigious Dillingham Commission on immigration wrote these distinctions into official government language in its influential report.

All these prejudices and theories became much more than mere verbiage when World War I broke out. Suddenly the restriction debates took a new and unexpected turn. The war effectively cut off immigration. It thus provided a kind of natural experiment for the employers' assertions that the economy required a continued large influx of new workers. Moreover, the doubtful loyalties of the immigrants to the foreign policy of their adopted country made many Americans wonder if the immigrant presence might not in fact constitute a threat to national security. For the restriction lobby, effective legislation began seem like an achievable goal.

World War I tore apart the domestic fabric of the country in a variety of ways. It unleashed an economic boom, as the United States supplied both agricultural products and manufactured goods to the warring nations—primarily the Allies. In 1917, in a move that was not popular among critical sectors of American opinion, the United States entered the war on the Allied side. Suddenly the pacifism of the Left, the pro-Central power leanings of the Germans, and the anti-English attitudes of the Irish became threats to national security. Prowar advocates, both within the government and among private lobbying groups, unleashed waves of official and vigilante violence against the dissenters in 1917 and 1918. Inflation and the end of the war in November 1918 produced a strike wave in 1919 and a sympathetic outpouring of feeling for the Russian Revolution. In late 1919 and 1920 the Justice Department conducted a Red Scare and rounded up hundreds of radical activists. Some were deported, some merely harassed. The infant American Communist Party felt the political climate in the nation was so hostile that it went underground (Higham, 1973; Murray, 1955).

In this atmosphere, immigration restriction became a mechanism for achieving domestic peace. As in the 1880s, immigrants, the restrictionists argued, were at the center of the labor violence and agitation. Open immigration was simply an invitation to admit Bolsheviks and anarchists into America.

Finally, the restrictionists argued, the war boom itself proved that the American industrial economy could survive and prosper without the continuing influx of European workers. Employers had successfully found other labor pools during war: rural Whites, Blacks from the South, women. One-half million southern Blacks, for example, had left the South for northern cities "during and shortly after World I" (Meier and Rudwick, 1970: 216), and proved themselves to be an acceptable substitute for immigrants. Black Americans spoke English, were willing to work, and were not Bolsheviks.

With the end of the war, mass immigration resumed. The literacy test was the chief restriction on entry and it proved to be a "coarse sieve" (Higham, 1973: 308). In 1921 Congress passed "emergency" restriction legislation, limiting the number of immigrants to 357,000 for one year. In line with the theories of the Immigration Commission, these slots were allocated to facilitate immigration from northern and western Europe and restrict the entry of southern and eastern Europeans. Each nationality was to receive an allocation based upon 3 percent of the strength of its population at the 1910 census. Since southern and eastern European immigrants were the latest arrivals, they had smaller quotas than those from nations that had been sending immigrants longer. The quota thus served as an objective mechanism of discrimination. It cut immigration from about 800,000 in 1921 to about 300,000 in 1922.

As a permanent measure, though, the law was problematic. It still admitted too many immigrants from southern and eastern Europe for the restrictionists. Further, the use of the census as the apportionment base was also suspect. The 1921 legislation had used the 1910 census because the 1920 data were not yet available. Logically, when these data became available, the apportionment base would shift to the newer data. But over time such a scheme would undercut the rationale for restriction because it would tend to increase the quotas for southern and eastern Europeans who were the most recent migrants and decrease them from for the northern and western Europeans who did not fill their quotas. The restrictionists suggested shifting the apportionment base to the 1890 census—before the start of the new immigrant wave—to avoid this problem, but Congress was leary of using such an openly discriminatory quota system. Fortunately for the restrictionists, the body of theory created by the racial thinkers came to the rescue. As I and others have shown elsewhere, the restrictionists were able to argue that the numbers in the 1890 census in fact approximated the "national origins" of the White population. Since the restrictionists wished to maintain the current "racial" character of the American White population, the National Origins Act of 1924 was the "fairest" measure of all. It severely restricted immigration from southern and eastern Europe and cut the overall annual quota to 150,000 (Conk, 1984; Higham, 1973).

This act ended the historic American open immigration policy. It also legitimated a set of racist theories about "higher" and "lower" grade White Europeans that suffused American social policy for the next forty years. In turn, such theories informed the American government policy to refuse to open the immigration door to European Jews persecuted by the Nazis during the 1930s and 1940s (Chase, 1977).

Ironically, though, the legislation did not end the migration of unskilled laborers to large American industrial cities. Rather, it encouraged employers to shift to other sources of labor—Blacks, rural Whites, and Hispanics. These groups were either already resident and citizens or, in the case of Hispanics from Latin America, outside the quotas of the National Origins Act.

Though some of the White migrants, for example, the Okies and Arkies of the 1930s and 1940s, captured public attention during their migrations, the most visible of these migrant streams have of course been the racial minorities who continue to be identified in census data. Fifteen million Blacks migrated during the 1920-1970 period, or around

300,000 a year. Like the European immigrants before them, Black migrants moved in waves. The two peak periods cluster around the world wars and hence the strong demand for labor in northern factories. This is not the place to enter into discussions about the comparative mobility rates or patterns of discrimination against Blacks and immigrants, but this voluminous literature does point to the basic similarities in the initial economic roles of the two migrant groups (Wilson, 1978; Lieberson, 1980; Pirore, 1979).

The patterns also hold for Hispanic migrants. In 1940, for example, there were about 70,000 Puerto Ricans in the continental United States. by 1970, there were 1.4 million. Of the total Puerto Rican population on the mainland in 1970, 59 percent resided in New York City (Maldonado-Denis, 1980: 131). In the Southwest, there have been significant migrations of Mexicans since the early part of the century. These immigrants worked as migratory agricultural labor in the corporate farms of the Southwest, and, after the 1940s, as an unskilled industrial labor force in the large cities. For example, there are currently over 2 million Hispanic residents of the Los Angeles-Long Beach metropolitan area, of whom 80 percent are Mexican American. Over 400,000 of these are Mexican immigrants who have come since 1970. Of the recent immigrants to the Los Angeles area, 55 percent work in unskilled blue-collar jobs; "half of all working Mexicans" were "concentrated in manufacturing" (Muller, 1984: 8; Acuna, 1972).

CONCLUSION

American immigration policy was formally governed by the 1924 National Origins Act until 1965. In reality, though, World War II, the Cold War, and the Black and Hispanic migrations began to undermine the logic of the law almost twenty years earlier. By the 1960s it was both an embarrassing and ineffective measure. Congress had passed so many exceptions and amendments that more immigrants were admitted outside the quotas than within them. During the 1950s, immigration was again running at the rate of 2.5 million a decade.

Several contradictory social and political tendencies converged to challenge the 1924 Act (Polenberg, 1980: 202-206). First, the racism inherent in the law was totally discredited by its obvious similarities to the fascist social theories of Hitler and Mussolini. The end of World War II revealed to Americans the true horror of Hitler's extermination policies for Jews and other "inferior" races. By the 1950s it was hard to use the racist rhetoric of the 1920s to classify the European population without treading dangerously close to fascism.

Further, the immigrants who wished to come to the United States from Southern and Eastern Europe after World War II were not degraded peasants, demented Bolsheviks, or Communist agitators. In fact, they were political refugees from the Communist nations of Eastern Europe or the displaced persons of World War II. They were a "better class" of people than the immigrants of the turn of the century; they quickly established themselves as loyal Americans, willing workers and strong anti-Communists. Again it was hard to justify the continued discrimination against persons coming from Southern and Eastern Europe.

Third, the National Origins Act had never governed immigration from the Americas—either Canada or Latin America. At the time the law was passed, immigration from the Western Hemisphere was both negligible and noncontroversial. Forty years later, this was patently not the case. Ironically, while the National Origins Act was undermined from one direction because it was discriminatory toward Southern and Eastern Europeans, it was undermined from another because it did not discriminate enough against Hispanic immigrants. Those who wished to extend the racially and ethnically discriminatory aspects of the law could agree that it missed the mark. The problem, though, was not the fact of discrimination. Rather, it was the groups discriminated against. Southern and western racists who wished to stem the Mexican immigrant tide demanded that quotas be created for the Americas.

Thus the liberals who wished to repeal the discriminatory legislation of the 1920s joined forces with the conservatives who wished to limit Latin American immigration to devise the Immigration Act of 1965. This law set quotas of 170,000 annually for immigrants from Europe, Asia, and Africa. No nation was to send more than 20,000 a year. It also extended the quotas to the Americas (120,000 a year), and set rules for preferential treatment of professionals and specially trained persons. Lyndon Johnson signed the law at the Statue of Liberty in October and claimed that American immigration policy was no longer based upon outdated and discredited racial policies. The nation admitted immigrants from all over the world.

Closer observers knew better. The rhetoric of antidiscrimination masked the flaws embedded in the legislation. The law generated two contradictory tendencies in the immigration patterns of the past twenty years. On the one hand, it encouraged a significant immigration of high skilled, professional and white-collar workers—often from developing

Asian nations. On the other hand it created a new pariah group in the United States—namely, the "undocumented" or "illegal aliens." The latter are now the subject of intense public debate.

The conservatives who wanted to limit Latin American immigrants did not anticipate that the simple creation of quotas would not deter massive immigration in the 1970s and 1980s. But American employers in crucial sectors of the economy—particularly low-skilled manufacturing and agribusiness—have succeeded in building complex systems for recruiting and employing cheap, pliant labor from Latin America. Since the early twentieth century, private contractors, the government sponsored bracero program, and the multicultural border towns have created networks that continue to attract migrants (Acuna, 1972; Garcia, 1980). Mexico and other Latin American nations have sharply growing populations and are themselves in the throes of industrialization and modernization. Agricultural workers are being displaced and forced into the urban labor market. Like the Polish, Russian, Italian, or Greek peasants of the turn of the century, these immigrants migrate to the cities with the most opportunity. At the turn of the century, the immigrants chose Warsaw, Moscow, Naples, and New York. Today, they might just as well go to Los Angeles or El Paso as to Mexico City. In fact, the wages are likely better in the United States.

By the late 1970s, the immigrant tide from Latin America and to a lesser extent Asia had reached a torrent. The government estimates that there approximately 6 million illegal immigrants in the United States currently; apprehensions of illegal immigrants along the borders exceed 1 million a year. In 1980, 125,000 "boat people" arrived from Cuba. The debt problems of Mexico further exacerbated the immigration situation; with the radical devaluation of the peso, the differential between the high wages of the United States and low wages at home further encouraged illegal entry.

The movement for immigration "reform" emerged early in the Carter administration, and led to congressional study of the situation and calls for both amnesty for illegals already in the country and sanctions against employers hiring illegal aliens. As during other debates about immigration restriction, there was a cacophony of voices. In the early 1980s, Hispanic organizations deplored the exploitation of the un-documented and called for legalization of their status. Labor unions called for punishment of employers who violated minimum wage and other labor laws. Growers employing large numbers of migrants de-manded a new bracero program. Local civic leaders called for reimburse-ment for the services they provided to the undocumented. And old-line conservatives warned ominously of the threat the immigrants posed to American values and culture. The rhetoric of reform began to speak of a racial "crisis" on our borders. For example, Clare Boothe Luce, former ambassador and currently a "special adviser" to the Reagan administration, told reporters for *Geo* in September 1982 that "cultural pluralism" could "become the enemy of the nation's will and purpose." She predicted "dreadful clashes" in the South and Southwest between the native and the foreign-born. These would be worse than those that occurred in earlier generations, she felt. The United States absorbed the earlier immigrants because "they were all white. They were not black or brown or yellow." The latter, she suggested, were like the "barbarians" of ancient Rome who turned the imperial center into a "city of pollution and noise and foreigners."

In such an atmosphere, Senator Alan Simpson and Representative Romano Mazzoli carefully began to work with the various interest groups to fashion a reform bill to end the flow of undocumented aliens and bring order to the immigration situation. As noted above, they came extremely close to success. They drafted a bill that melded the racist intent of some groups with the liberal goals of others. In the end they failed. As in the first two decades of the twentieth century, the current combined voices of the employers and immigrant communities are louder than those of the restrictionists. The racist rhetoric of restriction does not yet ring true to Americans. The country is prosperous, President Reagan tells us; the immigrant communities do not yet pose political threats to established elites. The extant mechanisms of restriction—border patrols and roundups—are cumbersome, expensive, and ineffective. The proposed new mechanisms of restriction—amnesty and employer sanctions—are even less palatable. So the amorphous status quo will continue to govern immigration for the near future. Whether the conditions supporting this situation will also continue remains to be seen.

2

Post-1965 Immigrants:
Demographic and Socioeconomic Profile

MORRISON G. WONG

☐ WE ARE A NATION of immigrants—some arriving before others, some arriving after others, but in the end, a nation of immigrants. Since the colonial period, four major waves of immigrants have arrived on our shores. However, because of this tremendous and sometimes sudden influx and some ethnocentric and racist ideologies, immigration policies sharply curtailed the flow of immigrants from about 1924 to about 1965. It is the most recent wave, since 1965, that is this chapter's concern. It is marked by the Immigration and Naturalization Act of 1965, the first real reform in U.S. policy in more than four decades.

The real effects of the 1965 act do not begin until 1968. In this year, one can see not only a resurgence in the number of immigrants, but also a dramatic change in their region of origin. These are significant changes; their implications and consequences have been the source of much discussion (Boyd, 1971, 1974; Keeley, 1971, 1974, 1975a, 1975b: North, 1974; Wong and Hirschman, 1983a).

This chapter has two objectives. First, it is to provide an overview and a better understanding of the changing trends and character of immigration to the United States since 1965. The second objective is to present an accurate and comprehensive portrayal, using recent data, of the new immigrants and their socioeconomic status in American society. The diversity of the immigrants from about 200 different countries makes it impossible to discuss each ethnic or nationality group separately. For the purposes of this chapter, the post-1965 immigrants will be divided by region of birth—Europe, Asia, North America, South

AUTHOR'S NOTE: *Special thanks to Bennett Fletcher of the Texas Christian University Computer Center for valuable time and expertise in data management and to James Henley for comments and critique of an earlier draft of this chapter.*

America, Africa, and Oceania. It is hoped that the insights and observations presented will point to some of the policy implications of this tremendous influx.

To accomplish this goal, the chapter is divided into five major sections. The first section presents the number and percentage of immigrants to the United States by region of birth for selected periods from 1961 to 1981, noting the impact of the 1965 Immigration and Naturalization Act. The second section presents demographic and social characteristics of these new (post-1965) immigrants by place of birth. In this section, as in subsequent ones, the new immigrants are compared with their native American counterparts. Third, the socio-economic characteristics of the new immigrants are addressed, in order to gauge their socioeconomic status as well as their contributions to the American economy. The fourth section considers the language and income assistance characteristics of the new immigrants in order to measure their social adjustment to American society. Finally, a discussion of the implications of these findings for policy formulation and implementation at the local and national level is presented. Areas for future research will also be discussed.

NUMBER AND PERCENTAGE OF IMMIGRANTS: 1961-1981

Before analyzing the social, demographic, economic, language, and income assistance characteristics of the new immigrants (post-1965), I should note the impact of the 1965 Immigration Act on the changing immigrant composition, as well as the changes in the region of origin. Table 2.1 presents data on the number and percentage of legal immigrants admitted to the United States by region of birth for selected periods from 1961 to 1981. For ease of analysis, these twenty years are divided into five major periods. The first period, from 1961 to 1965, marks the five years preceding the passage of the 1965 act. The second period, 1966 to 1968, is the transitional period. During this time, the backlog of immigrants under the old quota system was processed. The third period, 1969 to 1973, is the first five-year period after enactment of the 1965 act. In this period, one can note the impact of the 1965 act on the character as well as the region of origin of immigrants. The fourth and fifth periods, 1974 to 1977 and 1978 to 1981, suggest whether the initial impacts of the 1965 act represent an abnormal fluctuation or evidence of a particular immigration trend.

TABLE 2.1 Legal Immigrants Admitted to the United States, by Region of Birth

Region	Average Annual Number of Immigrants (in thousands)					Percentage Distribution for Each Period				
	1961-1965	1966-1968	1969-1973	1974-1977	1978-1981	1961-1965	1966-1968	1969-1973	1974-1977	1978-1981
Europe	122	133	103	79	68	42	35	27	18	12
Asia	22	54	104	152	235	8	14	28	35	43
North America	119	165	140	166	188	41	43	37	38	34
South America	24	21	21	27	38	8	6	6	6	7
Africa	3	4	7	8	13	1	1	2	2	2
Oceania	1	2	3	4	4	0	1	1	1	1
Total	290	380	377	436	547	100	100	100	100	100

SOURCE: U.S. Department of Justice (1981, 1975, 1974, 1973, 1972, 1970).
NOTE: The sums of the subtotals are slightly different from the total due to rounding error.

A glance at Table 2.1 reveals several interesting findings. First, the average annual number of legal immigrants increased almost twofold during this twenty-year period. It rose from an average annual number of about 290,000 in the 1961 to 1965 period to about 547,000 from 1978 to 1981. Indications as of 1984 suggest a continued increase in the average number of immigrants per year.

Second, there are significant changes in the region of origin. In the period before the 1965 Act, about 122,000 European immigrants arrived annually. However, in the last period (1978 to 1981), European immigration declined to about 68,000 per year. In other words, European immigration to the United States declined from 42 percent of the total immigrant population in the pre-1965 period to about 12 percent in the more recent period (1978 to 1981).

Third, during this twenty-year period, a dramatic shift occurred in Asian immigration, in numbers as well as in proportion to the total immigrant population. Pre-1965 Asian figures were at about 22,000 per year, or only about 8 percent of the total immigration stream. In the most recent period (1978 to 1981), the average annual Asian immigration rate jumped significantly to about 235,000 per year, accounting for about 43 percent of the total immigrant population, the highest proportion from any region.

Fourth, although immigration from North America increased from about 119,000 (pre-1965) to about 188,000 per year (1978 to 1981), this actually represents a proportionate decline of about 7 percent in the total immigration distribution. Nonetheless, legal immigrants from North America constituted a significant proportion of the total immigration stream throughout this twenty-year period.

Fifth, although the proportion of immigration from South America remained relatively constant at about 6 to 8 percent, this actually represents a slight numerical increase during this twenty-year period.

Finally, immigration from Africa and Oceania was relatively insignificant, accounting for about 1 to 3 percent of the total immigration population for any given period.

In sum, the major impacts of the 1965 act were (1) to increase the total number of legal immigrants for each successive time period; (2) to decrease the number and the proportional contribution of European immigration; and (3) to increase dramatically the number of Asian immigrants and their contribution to the immigrant stream.

With this historical analysis, we can now address the following questions: Who are these new immigrants? What are their social, demographic, economic, language, and income assistance characteristics? What are the implications of these findings for public policy formulation and implementation and for future research? The following sections provide some insight to these questions.

DATA AND VARIABLES

Before presenting the analysis of these characteristics of post-1965 immigrants by place of birth, it is useful to discuss the data source and describe the variables.

The major source of data is the 1980 U.S. Census Public Use Microdata Sample (PUMS) one-in-one thousand file. Because of the low number of immigrants from Africa and Oceania and the declining number of European immigrants, especially after 1965, all foreign-born individuals and a 2.5 percent sample of the native population were selected. This sample yielded 5270 native individuals and 7326 post-1965 immigrants. Of the immigrant population, there were 1231 individuals of European ancestry, 2244 Asians, 2,778 individuals from North America, 456 South Americans, 145 Africans, and 56 individuals from Oceania. The total for the subgroups differs from the grand total as 416 individuals were not classifiable for region of birth. (For more information, see U.S. Bureau of the Census, 1983.)

Most of the variables are self-explanatory. However, some require explanation. Sex ratio is a summary indicator of the sex composition of the population. It is the total number of males divided by the total number of females in the same population multiplied by 100. When the sex ratio is more than 100, males outnumber females.

Besides the categorical breakdown by specific age groups, the age composition of the population can be examined in terms of producers and dependents. To evaluate a subpopulation's dependency burden, three ratios are computed. The first is the youth dependency ratio (YDR). It is the total number of persons age 0 to 14 years divided by the total number of persons 15 to 64 years of age multiplied by 100. The second is the old age dependency ratio (OADR) and is computed from the total number of persons over age 65 divided by the total number of persons 15 to 64 years of age multiplied by 100. The last statistic is the total dependency ratio (TDR) and is calculated as a total number of persons 0 to 14 years and over 65 years of age divided by the total number of persons 15 to 64 years multiplied by 100. The total dependency ratio, then, gives us a rough estimate of how many unproductive dependents exist for every hundred persons of working age.

A measure of fertility is the child-woman ratio (CWR), which shows the number of children under age five per 1000 women of childbearing age (15 to 49 years of age) in a given year.

The labor force participation rate (LFPR) is a summary indicator of the proportion of the population involved in the labor force. It is calculated as the total number of persons 16 years and older in the labor force (whether employed or unemployed) and in the armed forces divided by the total number of persons over 16 years of age multiplied by 100. For more information, see U.S. Bureau of the Census (1983: K26).

Besides earnings or income, another measure of economic well-being is poverty status. The poverty status of a person who is a family member is determined by the family income and its relationship to the appropriate poverty threshold for that family. The poverty status of an unrelated individual is determined by his or her own income in relation to the appropriate threshold. Families or persons whose total family income or unrelated individual income in 1979 was less than the poverty threshold specified for the applicable family size, age of householder, and number of related children under 18 years present were in the category less than 1.00. Those above poverty level were classified into two groups, 1.00-1.99 and 2.00+, or those with income between 100 percent and 199 percent of poverty level, and those with income 200 or more percent of poverty level.

SOCIAL AND DEMOGRAPHIC CHARACTERISTICS

Table 2.2 presents the social and demographic characteristics of the native and post-1965 immigrant population by region of birth. Although space constraints do not permit elaboration on each group, several findings are important.

In general, the native and post-1965 immigrants tend to exhibit similar sex distributions. The only exceptions were the immigrants from South America (107) and especially from Africa (138), whose sex ratio was higher than the immigrant or native population, showing a greater proportion of males than females for each respective population.

Differences were noted in the age structure. In general, a greater proportion of the post-1965 immigrant population were between the ages of 19 to 34 years, or in the young adult category, than the native population. For all the other age categories, the native population was

TABLE 2.2 Social and Demographic Characteristics of the Native Population and Post-1965 Immigrants, by Region of Birth by Percentage

	Total Native Population	Total Post-1965 Immigrants	Post-1965 Immigrants					
			Europe	Asia	North America	South America	Africa	Oceania
Sex								
male	49	50	48	49	50	52	58	50
female	51	50	52	51	50	48	42	50
Sex ratio	95	100	94	95	101	107	138	100
Age								
0-18	31	26	24	26	25	24	19	38
19-34	27	42	34	43	44	44	54	38
35-64	30	29	37	28	28	29	27	25
65+	11	4	5	4	3	4	–	–
Youth dependency	37	23	19	25	21	17	17	33
Old age dependency	17	5	6	5	4	3	–	–
Total dependency	54	28	25	30	25	20	17	33
Marital Status								
single	43	42	35	44	41	40	33	48
separated-divorced	12	8	7	6	9	12	7	5
married	44	50	57	50	50	48	60	46

(continued)

TABLE 2.2 Continued

	Total Native Population	Total Post-1965 Immigrants	Post-1965 Immigrants					
			Europe	Asia	North America	South America	Africa	Oceania
Region of residence								
Northeast	21	28	46	21	20	52	34	20
North Central	27	13	18	16	9	5	11	5
South	34	21	17	24	27	23	32	5
West	18	38	9	37	44	21	23	70
Urban residence								
central city	18	45	36	45	48	50	41	55
SMSA-not central city	31	35	36	38	33	35	35	32
mixed-SMSA	30	16	23	13	15	12	23	11
not SMSA	21	4	5	4	4	3	1	2
Children								
none	34	27	33	26	24	25	26	23
0-6yrs	13	18	13	19	19	18	23	10
6-17yrs.	38	35	38	36	32	41	26	50
0-17yrs.	15	21	16	19	25	16	25	17
N	4573	6501	1112	1995	2510	409	119	250
Child/women ratio	288	94	69	103	82	90	111	52

SOURCE: 1980 U.S. Census Public Use Microdata Sample Tapes, One-in-a-Thousand Sample.

NOTE: Unless otherwise indicated the Ns are as follow: total native population = 5279; total post-1965 immigrants = 7326; Europe = 1231; Asia = 2244; North America = 2778; South America = 456; Africa = 145; and Oceania = 56. The sums of the subtotals are slightly different from the total due to rounding error.

larger. Most notably, about 11 percent of the native population was over 65 years of age. This is about three times higher than for the total immigrant population or for any of the immigrant groups. These findings are further supported by the high youth dependency ratio, old age dependency ratio, and total dependency ratio of the native population in comparison with their immigrant counterparts. A look at the mean age (not reported) shows that the post-1965 immigrants tend to be younger than the native population.

The post-1965 immigrants also are slightly more likely to be married and less likely to be divorced than the native although being married is the modal marital status for all immigrants (especially from Europe and Africa). One reason for a higher percentage of immigrants being in the married category is because of their proportion in prime marriage age groups. With few immigrant children and elderly, there are proportionally more immigrants in young and middle adult years.

Although post-1965 immigrants can be found in every part of the United States, they tend to concentrate in two major regions—the West and the Northeast. Immigrants seem to be quite selective. People from Europe, South America, and Africa tend toward the Northeast. Immigrants from Asia, North America, and Oceania tend to the West. Post-1965 immigrants live in four major states—California, New York, Texas, and Florida (Greenwood, 1983).

This residential concentration of the post-1965 immigrants is even more noteworthy when one recognizes that they are predominantly an urban population. Although there is some variation among the groups, about 45 percent live in the central city of a standard metropolitan statistical area (SMSA). An additional 35 percent live in an SMSA, but not in the central city. Hence, about 80 percent of the post-1965 immigrants are highly urbanized. This contrasts with only about 50 percent of the native population, who live in a highly urbanized area.

There are interesting differences in fertility. Perhaps most important, the new immigrants have more young children (0 to 6 years of age). The U.S.-born are more likely to have no children at all. But before we conclude that the post-1965 immigrants actually are more fertile than the U.S.-born, we should keep in mind the well-based suggestion that recent immigrants tend in time to match the general population in fertility (Bouvier, 1983: 197). The child-to-woman ratio (CWR) presents a slightly different picture. This ratio for the native population is 288, or 288 children under age 5 years per 1000 native-born women of child-bearing age. This ratio is about three times greater than the post-1965 immigrant population figure, suggesting that native women are having

TABLE 2.3 Socioeconomic Characteristics of the Native Population and Post-1965 Immigrants, by Region of Birth by Percentage

	Native Population	Post-1965 Immigrants	Europe	Asia	North America	South America	Africa	Oceania
Education								
never attended	1	4	4	4	5	1	1	–
elementary	6	20	16	9	33	12	6	7
some high school	25	18	19	12	23	23	2	17
high school graduate	38	22	29	20	21	24	16	28
some college	14	14	13	16	10	19	38	24
college graduate	9	9	7	17	3	5	10	14
post graduate	7	12	12	21	4	15	27	10
Mean years completed	11.8	10.9	11.11	12.9	8.8	11.8	14.3	12.8
N	3041	4409	815	1376	1608	279	96	29
Occupation[a]								
executive-manager	11	8	9	9	4	8	15	12
professional	11	13	14	20	6	14	22	19
technical	3	3	7	5	16	3	4	4
sales	10	6	7	7	4	6	4	–
administrative support	17	13	11	16	10	13	14	12
service	13	17	15	15	20	16	15	27
farm	14	3	1	1	6	1	–	–

crafts	14	12	14	8	14	14	10	12
operatives	18	26	26	19	34	27	18	15
N	2802	3474	657	1068	1272	222	74	26
Industry[a]								
agriculture	11	9	10	4	14	3	4	–
manufacturing	23	30	33	28	31	36	15	15
transport	8	4	4	4	5	5	7	4
wholesale trade	4	4	4	4	4	4	3	4
retail trade	14	15	16	17	13	10	12	15
finance	6	5	5	6	4	6	10	–
business services	4	5	4	5	6	9	10	4
personal services	4	6	6	5	8	5	4	8
professional services	20	19	15	25	14	20	34	38
public administration	5	3	2	4	2	2	3	12
N	2202	3474	657	1068	1272	222	74	26
Class of worker[a]								
private	70	81	84	74	88	78	72	67
government	18	11	8	16	7	11	18	33
self-employed	11	8	9	9	5	11	11	–
family	1	1	5	1	1	–	–	–
N	2222	3514	659	1092	1279	226	74	27

(continued)

TABLE 2.3 Continued

	Native Population	Post-1965 Immigrants	Europe	Asia	North America	South America	Africa	Oceania
Labor force participation[a]	62	70	70	70	72	69	69	76
N	3044	4409	815	1376	1608	279	96	29
Income[b] family	23.5	21.2	24.9	24.4	17.1	20.8	21.6	26.4
N	4547	6319	1107	1908	2456	401	110	52
wages & salaries[c]	17.7	14.4	16.7	15.8	11.8	14.7	16.8	17.0
N	1015	1469	288	442	559	75	26	9
Total[c]	18.9	15.3	18.0	16.9	12.0	16.6	21.5	18.1
N	1112	1546	306	470	571	89	29	9

SOURCE: See Table 2.2.

NOTE: The sums of the subtotals are slightly different from the total due to rounding error.

a. Ages 25 and over.

b. In hundreds of dollars.

c. Ages 25 and over, with an occupation, and working full time with positive earnings.

children closer together in time than their immigrant counterparts. However, probably much of these differences is explained by the fact that these ratios are simplistic and do not take into account the various ways in which these groups differ from each other demographically.

SOCIOECONOMIC CHARACTERISTICS

We can now investigate the socioeconomic characteristics of the post-1965 immigrants by region of birth. How do they fare in comparison with their native-born counterparts?

A glance at Table 2.3 reveals several general findings. In terms of education, the post-1965 immigrants are overrepresented in elementary education, underrepresented at the middle level (high school), and equal at the college level. Except for immigrants from North America, they are overrepresented at the post graduate level. It is interesting that about 21 percent of the Asian immigrants and 27 percent of the Africans (compared to only 7 percent of the native population, age 25 and over), have some post graduate education. Although the mean number of years of schooling completed of the post-1965 immigrants is about one year below the native average, immigrants from Asia, Africa, and Oceania completed more years of education than did their native counterparts.

There are many differences in the occupational distribution for the adults, but they tend to be minor. One exception is the slightly greater appearance of immigrants than native as operatives (26 percent and 18 percent). Yet, it is worth noting that the total occupational distribution of all immigrants tends to mask differences by region of birth. Immigrants from Europe, North America, and South America tend to be concentrated in operative occupations; Asians in professions and operatives; Africans in professions; and Oceanic immigrants in service occupations. Separate analyses for males and females for each immigrant group suggest only slight variations in the occupational distribution: Males are overrepresented in executive-manager and crafts occupations and females in administrative support and service occupations. (Because of the small number of immigrants from Oceania, any findings regarding this group are, at best, tentative.) The industrial distribution for native and immigrant adults is very similar. The only exception is in the industrial category of manufacturing where post-1965 immigrants are overrepresented against the native (30 percent and 23 percent, respectively). The two major areas of concentration were in manufacturing and professional services. This was the case regardless of nativity status or region of birth. Due to the residential concentration of post-1965 immigrants, differences in occupational and industrial distribution may be a reflection of regional or urban effects.

Over two-thirds of both the native and post-1965 immigrant adults work in the private sector of the economy, with immigrants being more represented. As expected, citizenship restrictions on most government jobs means that post-1965 immigrants tend to be slightly less represented in the government sector. One would think that with the expansion of the various ethnic enclaves and the proliferation of ethnic businesses (restaurants, mom-and-pop grocery stores, and assorted specialty shops), a greater proportion of the immigrant population would be self-employed and involved as family workers. However, the data suggest the reverse; a greater proportion of the native were self-employed. Moreover, there was no difference in the proportion of native or post-1965 immigrants who classified themselves as family workers. Finally, the labor force participation rate of the post-1965 immigrants was about 7 to 14 percent higher than the native-born population.

Our last socioeconomic characteristic of the post-1965 immigrants is income. In order to gain a more comprehensive picture, three separate measures of income were analyzed. The first measure is annual family income. In general, post-1965 immigrant families earn about $21,200— about $2300 less than the mean native family income. However, by looking at the origins of immigrants, we see a different pattern of earnings. Immigrants from Europe, Asia, and Oceania have higher family earnings than the native. One possible explanation might be multiple jobholders in the family. On the other hand, immigrants from North America drew the lowest mean family income: about $4100 less than the post-1965 immigrant average and about $5400 less than the native average.

A second measure of income is wages and/or salaries for adults working full time. By this measure, the mean wages and/or salary of native individuals were about $17,700 or about $3300 more than post-1965 immigrants. In no case did this measure show any immigrant group having a higher average than the native. North Americans were the only group that had mean wages and/or salaries below the total immigrant level.

The last measure of income analyzed was total income. This includes income not only from salary and/or wages, but also from self-employment, interest and dividends, social security, public assistance, and all other sources. In general, post-1965 immigrants earn about

$3600 less in total income than the native population. Again, immigrants from North America had the dubious distinction of being the only immigrant group earning less than the total immigrant average.

In sum, by any measure, post-1965 immigrants earned less than their native counterparts and North Americans earned the least of all groups. Chiswick (1978) has reported, however, that White male immigrants reach earnings parity with the native after about thirteen years and earn more in subsequent years. Whether this parity will be reached by non-White immigrants remains an open question.

LANGUAGE AND ASSISTANCE CHARACTERISTICS

This section presents a discussion of more general social factors: the language and income assistance characteristics of the native population and post-1965 immigrants.

We see that nearly all of the native population speak English at home; less than one of each five post-1965 immigrants does. That is, better than 90 percent of the later immigrants speak a foreign language at home (see Table 2.4). Home use of a foreign language does not necessarily mean that the individual cannot or does not speak English; possibly the individual is bilingual. The data show this. Of the post-1965 arrivals who speak a language other than English at home, about 40 percent either do not speak English or speak it poorly. This contrasts with the only 8 percent of the native people who responded in like fashion. North Americans (mostly Hispanics) are the most handicapped; approximately 30 percent do not speak English very well. An additional 25 percent do not speak English at all. Except for Africans, about 30 percent of the other immigrant groups are not fluent in English.

We can look at the economic assistance required by the post-1965 immigrant and native population. Data on two types of economic assistance are presented: social security and public assistance.

The native population receives about $700 more in social security payments than the post-1965 immigrants. Moreover, a greater proportion of the native than the immigrant population receive social security payments (13 percent and 2 percent, respectively).

Opposite findings are noted for the distribution of public assistance. Post-1965 immigrants receive about $355 more than the native population. Asian immigrants receive about $2627, the highest public assistance of any group. A partial explanation for these high payments may be the government aid that many Vietnamese refugees receive to

TABLE 2.4 Language and Assistance Characteristics of the Native Population and Post-1965 Immigrants by Region of Birth by Percentage

	Native Population	Post-1965 Immigrants	Europe	Asia	North America	South America	Africa	Oceania
Language spoken								
Foreign	7	82	74	91	83	82	79	49
English	93	18	26	9	17	18	21	51
N	5039	7196	1216	2203	2734	449	143	53
Facility with English								
very well	70	33	43	39	22	38	63	62
well	22	28	25	34	23	31	26	8
not well	6	25	24	20	30	23	9	31
not at all	2	15	9	7	25	8	3	–
N	339	5910	902	2005	2275	369	113	26
Income								
Social Security	3067	2346	2378	2569	2251	4335	35	875
N	660	167	54	30	58	2	2	1
Public Assistance	2065	2420	2135	2627	2314	2635	2742	725
N	178	272	38	70	121	18	3	2
Percentage of population	3	4	3	3	4	4	2	4
Poverty Status								
less than 1.00	12	20	9	18	25	16	26	20
1.00 - 1.99	19	26	20	21	33	30	22	6
2.00+	69	54	72	61	42	54	52	74
N	5127	7194	1224	2203	2740	440	141	54

SOURCE: See Table 2.2.
NOTE: The sums of the subtotals are slightly different from the total due to rounding error.

ease their adjustment. North (1983) found that about 80 percent of the Indochinese refugees receive Refugee Cash Assistance (RCA), food stamps, and/or Medicare. However, one should not conclude that the immigrant groups are bleeding the federal budget. Only about 3 percent to 4 percent of the immigrant populations receive public assistance monies. Moreover, their small proportion relative to the total U.S. population must be considered.

Our final indicator of economic well-being is poverty status. About 20 percent of the post-1965 immigrants live at or below poverty level. This figure is about 8 percentage points higher than the 12 percent figure for the native population. More specifically, about one-quarter of the immigrants from North America and from Africa live at or below poverty level. This latter finding is interesting considering the above average income level of the African immigrants, suggesting a bimodal income distribution. However, the figures also show that the majority of all immigrant groups, with the exception of those from North America, are at least once removed from the poverty line.

IMPLICATIONS FOR PUBLIC POLICY AND RESEARCH

With this overview of the social, demographic, economic, language, and income assistance characteristics of the post-1965 immigrants to the United States, we can now ask: What are the implications of this new and tremendous influx for American society? What can be extrapolated to help in policy formulation and its implementation? What are some of the unresolved questions for further research in order to better understand the immigration process and socioeconomic status of the immigrants in the United States?

Briefly, one of the most significant changes of the 1965 Immigration Act has been the tremendous increase (almost double) in the number of legal immigrants per year since the early 1960s. Another significant change is the shift from Europe to Asia. Currently, immigration from Asia accounts for about 43 percent of the total immigration population each year. Recent patterns suggest that the proportion of new Asian arrivals will continue to increase. Although the proportion of immigrants from North America (particularly Mexico) has remained relatively constant, it has increased slightly in numbers during this twenty year period. Hence, legal North American immigrants still are about one-third of the yearly arrivals. Thus, about 75 percent of all immigrants currently come from only two regions—Asia and North America. If current immigration trends persist, the social character of American society surely will change. The United States may be entering a new phase of heterogeneity—from a multiethnic to a multiracial society (Bouvier, 1983: 197).

Various ethnic communities and enclaves in major urban areas have expanded and others are being established as American society and the various indigenous ethnic subcultures attempt to absorb the new immigrants. Moreover, the ideological orientation with regard to the future of race and ethnic relations in American society is sure to change. Although many Americans have given lip service to the idea of pluralism, the influx of huge numbers of immigrants from Third World countries who exhibit considerable ethnic pride and a desire a maintain their ethnic identity may be the catalyst for more serious discussion on the plausibility and advantages of a pluralistic orientation, and less emphasis on an assimilation or Anglo-conformity orientation (Fuchs, 1983). This recent influx from Asia and North America is not without its consequences. There currently are indications that ethnic antagonism, particularly directed at Asians and Hispanics, is on the rise. This is particularly noteworthy in the major urban areas with a large concentration of immigrants.

Because about three-quarters of the new arrivals each year come from Asia and North America, the specific impact of these two groups on their new locales must be discussed. Post-1965 immigrants from Asia and from North America are residentially concentrated. They have tended to settle in the western and, to a slightly lesser degree, in southwestern states (Greenwood, 1983). Furthermore, over 80 percent live in highly urbanized areas. Policymakers, particularly those in urban areas of highly concentrated Asian and Mexican immigrants, must be aware of the tremendous impact this influx has upon social, educational, economic, and related institutions and resources. Young immigrants will greatly increase educational enrollments. A lack of facility with English, more marked among immigrants from Mexico, is a cause for serious discussion regarding bilingual/bicultural programs in the educational system, as well as hiring bilingual staff in the service agencies. How can immigrants receive adequate health care if no doctors, nurses, or hospital personnel speak the immigrant's language? Are immigrants receiving equal protection under the law if neither police nor judges can communicate? Can they be served equitably in other ways?

On most socioeconomic indicators, post-1965 Asian immigrants have higher achievement than the native population. A greater proportion of post-1965 Asian immigrants have graduated from college and are involved in the professional occupations or industries than the native or general immigrant population (Hirschman and Wong, 1983). The average family income of the Asian immigrants is also higher than the native and post-1965 immigrant average, but this may be partially a reflection of the presence of multiple earners in the family (Wong, 1983; Wong and Hirschman, 1983b). The wages and/or salaries and total income for the Asian population was higher than the immigrant average, but lower than the native average. This finding is disturbing in that a greater proportion of Asian immigrants than native are in the professional occupations or industries which are not only higher in status, but also higher in economic returns. At the opposite end, about 18 percent of the Asian population were at or below the poverty line. This percentage is lower than the general immigrant population, but higher than the native population. In sum, although post-1965 Asian immigrants have achieved considerable socioeconomic status, this has been more in terms of education and occupation. Their economic returns are still deficient when compared to the native population. This conclusion substantiates the findings of past research on Asian socioeconomic achievement (Bach and Bach, 1980; Chiswick, 1983; Marsh, 1980; Massey, 1981; Montero, 1979; Wong, 1980; 1982).

The post-1965 North American immigrants, particularly those from Mexico, did not fare as well as their Asian counterparts. Immigrants from North America are disadvantaged in terms of their education, their proportion in professional occupations, and on all three indicators of income—family, wages, and total income compared to other immigrants and the native population. About one-fourth of the North American immigrants are living at or below the poverty line. This is about five percentage points higher than the immigrant average and about twice the native average.

In sum, it seems that for these two major post-1965 immigrant groups, there are two almost completely different stories of socioeconomic achievement. For the post-1965 Asian immigrant population, the story seems to be quite rosy. There seems to be very few barriers to socioeconomic achievement, having higher achievements on most indicators than the post-1965 immigrant and the native population. Asian immigrants seem to be disadvantaged only on income. Explanations for the lower economic returns for educational and occupational characteristics of the Asians should be a focus of future research (Jiobu, 1976; Wong, 1983). For the post-1965 North American immigrants, the story of their socioeconomic achievement is one of disadvantage. On most socioeconomic indicators, their achievements are not only lower than the native population, but also lower than the general immigrant population. Future research should explore the possible explanations that account for the disadvantage of the North American immigrants to the United States, looking at their social, demographic, occupational, and income characteristics (Borjas, 1981; Portes and Bach, 1980; Poston and Alvirez, 1973; Poston et al., 1976; Reimers, 1980; Tienda, 1981, 1983). Future research may shed some light on the differential degree of socioeconomic achievement as well as what factors best explain this situation (Chiswick, 1978, 1979, 1980; Tienda, 1983). Moreover, many questions regarding the entry of immigrants into the labor market need further study (Cornelius, 1979; Greenwood, 1983; Killingsworth, 1983). Such questions may include the following: (1) Are immigrants displacing the native population occupationally? (2) How do immigrants create jobs or, alternatively, take away jobs from native Americans? (3) Do immigrants affect the prevailing wage structure in the various locations (regional or urban areas) in which they reside? (4) Who benefits and who is deprived by an influx of immigrants into a certain locale?

This study is able to analyze only briefly the most general characteristics of the post-1965 immigrants. Future researchers, using other data sources, may wish to extend this analysis of the socioeconomic characteristics of these immigrants. Furthermore, although there are some similarities among the various immigrant groups to justify analyzing their characteristics by region of birth, one should be aware that there are also tremendous differences between the various immigrant groups within each regional grouping. Although it may be informative to discuss post-1965 Asian immigrants in general, there are tremendous differences among the Chinese, Japanese, Filipino, Korean, Vietnamese, and other Asian immigrant groups, not only in their social, demographic, and socioeconomic characteristics, but also in the customs, residential patterns, language patterns, and general socioeconomic characteristics and achievement patterns (Hirschman and Wong, 1981; Wong and Hirschman 1982). Placing these various groups into one category of "Asian immigrants" masks their diversity as well as complicates analyzing and explaining their socioeconomic achievement. This argument also holds true for other regional immigrant groups such as North Americans who can be analyzed separately as Mexicans, Cubans, or Dominicans. Future research may wish to analyze, in greater depth and with more sophisticated statistical methods, the social and economic characteristics of each of the major immigrant groups to

American society (Chiswick, 1979; 1983; Hirschman and Wong, 1981; 1984; Wong, 1982).

Before ending, there are two aspects of immigration to the United States that should be mentioned. First, it should be recognized that some of these immigrants, particularly those from Southeast Asia and Cuba, came to the United States under special circumstances, as refugees. As a consequence, the social, economic, and political circumstances surrounding their entry to the United States were different than for most other immigrants. Hence, for these refugees-turned-immigrants, much of their immigration in the 1970s, as well as their socioeconomic achievements, may represent an anomaly—a situation not destined to repeat itself in the near future. Second, the focus of this chapter has been on legal immigration to the United States. Anywhere from 2 million to 6 million undocumented aliens are currently living in the United States. Their impact on American society is significant. Unfortunately, with only limited reliable data sources and because of the illegal nature of their entry, empirical research on their socioeconomic status and achievements as well as their contribution to the economy and society is scarce and sometimes contradictory. Thus this chapter dealt only with legal immigration to the United States. Future research should focus on the impact of the undocumented aliens.

3

Immigration Issues in Urban Ecology: The Case of Los Angeles

PHILIP GARCIA

☐ OUR NEED FOR INFORMATION on the most recent immigrants from Latin America, Southeast Asia, and other non-European nations spans many research topics. The findings presented here are a response to that need and will provide an update on the status of immigrant colonies and ethnic ghettos in urban centers. My purpose is twofold. The main task is to evaluate the current settlement patterns of some of the new urban immigrants and detect any trends in their geographical dispersion after arrival. I will focus on the degree to which traditional ecological concepts such as residential invasion and succession explain the spatial distribution of the new immigrants. Second, I will present some preliminary remarks on how their settlement patterns affect their relationship with traditional minority and majority group members. The main concern of the chapter is what these relationships imply for the groups' economic status, cohesion, and solidarity.

SCOPE OF WORK AND METHODS

During the 1970s more than 5 million immigrants entered the United States; they settled in every major region of the country. Of course, some states received a disproportionate share. The five states with the largest proportion of new entrants were California, New York, Texas, Illinois, and Florida (Passel and Woodrow, 1984). Obviously, an examination of residential segregation for a national sample of urban areas is an important research task, but it is not the route followed here. Instead, this chapter takes a case study approach.

The target is Los Angeles County. This civic division is the areal equivalent of the Los Angeles-Long Beach Standard Metropolitan

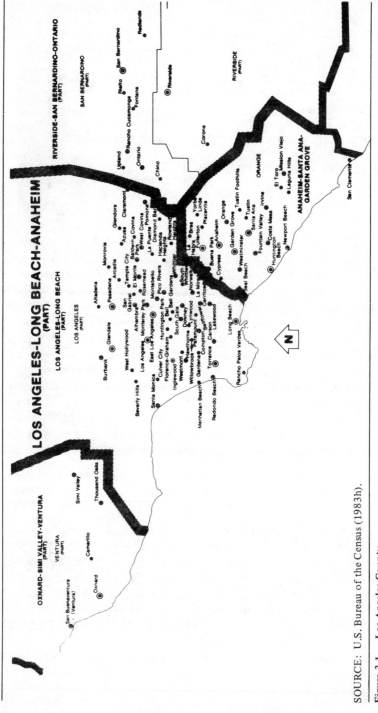

SOURCE: U.S. Bureau of the Census (1983h).

Figure 3.1   Los Angeles County

SOURCE: L.A. City Planning Division.

Figure 3.2   City of Los Angeles

Statistical Area. Los Angeles County has been a port of entry for many post-World War II immigrants. Close to a quarter of its population is now foreign-born. Almost three-quarters of Los Angeles's foreign-born arrived between 1965 and 1980. It has sizable populations of native American Hispanics, Asians, and Blacks (see Appendix Tables 3.A2 and 3.A3).

Data. The demographic analyses are based on enumerations from the 1980 census. The basic unit is the census tract. For Los Angeles, the average tract contains about 4500 residents. Tracts are separated into those within the boundaries of the city and those outside the municipality (see Figures 3.1 and 3.2).

The ethnic and racial categories used here are as follows: Hispanics, Whites or Anglos, Blacks, Asians, and American Indians. Unfortunately, tract data do not disaggregate each ethnic/racial group by nativity. The data do allow us to partition the total foreign-born population into two groups based on their year of immigration. Those who arrived between 1965 and 1980 are referred to as "new" immigrants and those who arrived before 1965 are called "old" immigrants.

Measures. Residential segregation has many faces. Some areas have large proportions of minority residents; others do not. Some are characterized by the presence of two major ethnic/racial groups, whereas others have many. Which community would require the most population movement to achieve desegregation: one with a small percentage of minority group members who are highly segregated or one with a large minority population moderately segregated? Or, how does one contrast a biethnic neighborhood with a multiethnic neighborhood in terms of interethnic variety, given that population sizes frequently differ? To answer such questions one must rely on several statistics.

Besides population percentages, three more or less common demographic measures are used here that, taken together, describe various properties of ethnic/racial settlement and segregation. They are the index of dissimilarity (ID), a variant of the replacement index (RI), and the heterogeneity index (HI).

The ID is the primary segregation index. It refers to the percentage of one population that would have to move in order to achieve an even distribution throughout a geographical area. A score of 100 indicates complete segregation between two groups; zero indicates complete integration.

The RI is also a segregation index, but it equals the minimum percentage of the combined population of two groups that would have to move so that each subpopulation could achieve an even distribution throughout a given area. The RI has the same range of possible scores as the ID, and the scores carry the same basic interpretation: The higher the score, the greater the amount of population movement required to desegregate two groups. If we multiply the RI by the number of persons in each of the two ethnic/racial categories, we have an estimate of the actual number of persons who would have to exchange their places of residence in order to achieve complete desegregation.

The third index (HI) summarizes the level of ethnic/racial diversity found in a community. The higher the HI score, the more likely residents of one group will live next door to members of another group. The index takes into account the number of groups present in an area and their relative sizes. The specifics of all three measures are described in Appendix Table 3.A1.

CHANGES IN ETHNIC COMPOSITION FOR LOS ANGELES

Proportionally, Los Angeles County may have received more international migrants between 1965 and 1980 than any other metropolitan area in the country. Published 1980 census figures suggest that 15.6 percent of its total population consisted of immigrants who arrived after 1965. By contrast, the comparable proportion for the New York urban area, the second largest metropolitan recipient, was 10.2 percent.

This level of immigration has certainly increased the saliency of immigrant status in Los Angeles County. Between 1950 and 1980 the percentage of the foreign-born residents rose from 10 percent to 22 percent. Published figures also indicate that the current immigrant population is overwhelmingly Hispanic and Asian.

About half of all Hispanics and almost two-thirds of all Asians were born in other countries. The other three groups are essentially native populations. These proportions relieve some apprehensions when we are forced to use Hispanic and Asian figures to reveal the probable trend in residential dispersion among the new immigrants (see Table 3.1).

It is not too surprising that most contemporary discussions of immigration or ethnicity in Los Angeles County downplay the issue of European nativity or parentage. Only about 13 percent of its foreign-born residents came from European nations; fewer than 5 percent of all immigrants who entered since 1965 reported a European heritage.

In addition to the small numbers of new European entrants, other factors help to make individual European ancestry groups socially inconspicuous. For example, there does not seem to be any group of

TABLE 3.1   Percentage Distribution of the Foreign-Born, by Ethnic
and Racial Status, for Los Angeles County: 1980

Year of Immigration	Total	Anglo	Spanish	Black	Asian	Indian
1965-1980	7.7	5.2	34.1	0.9	27.5	2.0
Before 1965	14.9	5.8	11.5	1.9	35.1	4.4
Native	74.4	89.0	54.4	97.2	37.4	93.4
	100.0	100.0	100.0	100.0	100.0	100.0
Foreign-born	22.6	11.0	45.6	2.8	62.6	6.6

SOURCE: U.S. Bureau of the Census (1983c).

European ethnics with a large proportion of non-English-speaking monolinguals or English-speaking bilinguals. And European ethnics in California appear to be quite homogeneous across the gamut of socioeconomic indicators provided by the Census Bureau.

The absence of other cultural differences no doubt further reduces the social significance of European backgrounds. For instance, among the four largest European ancestry groups (English, German, Irish, and Italian), mixed ancestry is very common. As a consequence, spatial isolation for these groups from other Anglos looks to be very uncommon (see Table 3.2).

If there is an Anglo subgroup with high visibility, it is the Jewish community. Along with its Mexican Plaza, Chinatown, Little Tokyo, and Koreatown districts, Los Angeles has also a traditional Jewish storefront area in the west-Wilshire district. Jewish neighborhoods were located in the center of the city of Los Angeles as early as the teens (Romo, 1982). Their highest residential densities are now in the west and northwest parts of the city.

Because the Census Bureau does not enumerate religious affiliations, it is very difficult to determine the population parameters of the Jewish community or its level of spatial isolation. Since 1950, the number of American Jews in the area has been estimated at about 500,000.

Traditionally, it has been assumed that people of Russian descent were Jewish (see Rosenthal, 1975). If Russian ancestry is still a reasonable surrogate for American Jews, then the Los Angeles Jewish population is more segregated from other Anglos than the Anglo ethnics listed in Table 3.2. This is because the Russian ancestry group has a much higher percentage within the city than in the county as a whole. To desegregate within the city, 34.4 percent of the Russians would have to move to Anglo neighborhoods.

TABLE 3.2 Social Characteristics of Selected Anglos in Los Angeles County Who Self-Reported European Ancestry: 1980

Characteristic	English	German	Irish	Italian	All Other Anglos
Number	1,177,674	983,127	73,865	289,676	1,629,261
Percentage of total population	15.6	13.1	11.7	3.8	21.8
Percentage reporting multiple ancestries	62.6	71.1	79.5	47.2	—
Index of dissimilarity	21.7	17.0	15.6	17.7	—
Replacement index	5.7	3.9	3.2	1.3	—

SOURCE: Garcia (1985).

NOTE: The contrast group for the index of dissimilarity was the sum of all other Anglos.

TABLE 3.3 Ethnic and Racial Characteristics for Los Angeles County: 1950-1980 (in thousands)

	1950		1960		1970		1980	
	Number	%	Number	%	Number	%	Number	%
Anglo	3,582.9	86.3	4,770.7	79.0	4,640.7	65.9	3,978.0	53.2
Hispanic	286.5	6.9	597.8	9.9	1,093.0	18.1	2,063.8	27.6
Black	220.0	5.3	531.4	8.8	724.7	12.0	942.2	12.6
Asian[a]	62.3	1.5	138.9	2.3	187.2	3.1	478.6	6.4
Total	4,151.7	100.0	6,038.8	100.0	7,042.0	100.0	7,477.5	100.0
Heterogeneity index		24.8		35.8		51.2		62.1

Decennial Changes

	1950-1960		1960-1970		1970-1980		1950-1980	
	Number	% Change	Number	% Change	Number	% Change	Number	% Change
Anglo	1,187.8	33.2	-130.0	-2.3	-662.7	-14.3	395.1	11.0
Hispanic	311.3	108.7	495.2	82.8	970.8	88.8	1,777.3	620.3
Black	311.4	141.5	193.3	36.4	217.5	30.0	722.2	328.3
Asian[a]	76.6	123.0	48.3	34.8	291.4	155.0	416.3	668.2
Total	1,887.1	45.5	1,003.2	16.6	435.5	6.2	3,325.8	80.1

SOURCE: Community Analysis Bureau (1977), U.S. Bureau of the Census (1983c).
a. Contains a small number of American Indian and other ethnic enumerations.

Without a doubt, the predominance of non-Anglo immigration has helped to reshape the general perception of what is ethnic and dramatically alter the ethnic/racial landscape of Southern California. Compared to the year 1980, the immediate postwar population was relatively homogeneous with regard to the four ethnic/racial categories. In 1950, for instance, the likelihood, in percentages, that two randomly selected residents were of different ethnic/racial background was only about 25 percent. In 1980, the odds were close to 62 percent.

Table 3.3 shows the absolute and proportional increases for each ethnic/racial population between 1950 and 1980. Though Anglo is still the largest category, the countywide percentage declined by one-third during the three decades. Moreover, unlike any of the other ethnic/racial groups, the actual population size of Anglos has been decreasing.

Clearly, Hispanics have contributed most to the postwar population growth of Los Angeles, and they have been the most numerous minority group. (Because of their continuously high growth rate, Hispanics may be at their peak in terms of social visibility.) But as the lower panel of Table 3.3 suggests, each non-Anglo group appears to have had its decade in terms of experiencing the highest proportional rate of net growth: The 1950s belonged to Blacks, the 1960s to Hispanics, and the 1970s to Asians.

Generally speaking, ethnic/racial residential patterns have followed the changing ethnic/racial composition of the population. The number of tracts with overrepresentation of Anglos is decreasing, while those dominated by non-Anglos are increasing. To highlight some specifics on whether the latest ethnic/racial changes have meshed with the overall pattern of population redistribution, we follow an agenda of questions closely derived from the human ecology literature.

IMMIGRATION AND URBAN ECOLOGY

The topic of immigrant settlement in North American cities carries with it a long history of research. The recorded experiences of the millions of European immigrants who arrived on the eastern shores between the Civil War and World War I are linked inseparably with the onset and development of human ecology as a sociological study area. Data on those second wave immigrants were a cornerstone of the urban ecological structure in the United States. It was observed that low status groups are the primary residents of areas located in the center of the city, whereas high status groups reside outside the inner city. Such data were the heart of the descriptions that outlined the stages of residential change that occur within cities (see Theodorson, 1961).

Over the years the many propositions associated with the early human ecologists were used to account for the residential redistribution of southern Blacks to nonsouthern cities and other immigrants to various metropolitan areas. The findings were mixed. Some researchers, for example, found that distance from the core of an urban area could account for gross residential patterns among status groups. But more of the residential variation could be explained by sectoral development (e.g., Van Arsdol and Schuerman, 1971). Such discordant results were used by critics to counter ecological determinism. In regard to the urban settlement of more recent migrants, the residential patterns of southern-born Blacks seldom conformed to the patterns of their European predecessors. Here, the recurring challenge by critics is whether the ecological models based on second wave immigrants can be generalized to later waves of immigrants who may have arrived under different national and international circumstances.

Nevertheless, there has been fairly consistent support for more than a few of the original observations made by the early human ecologists. Two of the more stable characteristics of U.S. cities have been the spatial clustering of similar socioeconomic status groups and the residential segregation of ethnic/racial minorities. Perhaps because of the enduring nature of these traits, researchers have found it profitable to reexamine the classical propositions of human ecology periodically with new data.

ARE IMMIGRANTS IN THE INNER CITY?

Several old research questions seem relevant to our interests. Do immigrants still make their initial entry in those residential areas where housing is most available, that is, in the oldest parts of the city?

Immigrants traditionally have been located near a city's downtown area, as in Chicago. The conventional scenario is that new arrivals flock to these older areas (where other immigrants probably first settled); their entry usually means that older residents move to other neighborhoods. Over time, the new immigrants and their offspring become the dominant residents of the older residential areas. In other words, entry communities of first settlement are subject to invasion and succession. If this is the current pattern in Los Angeles, new arrivals would be overrepresented in residential areas nearest the inner city, and they would be more residentially segregated from Anglos than from older immigrant waves.

The traditional central business district or downtown area of Los Angeles is surrounded by seven communities that were established prior to the 1920s: Boyle Heights, Central, Elysian Park, Lincoln Heights, University, Westlake, and the Wholesale District (see Figure 3.2). Approximately 400,000 people reside in these centrally located neighborhoods. Nearly 50 percent of the population in these communities is foreign-born, or double the countywide percentage. New immigrants are more than twice as likely to be found in this central area as in the rest of the county. This overrepresentation is reflected in ethnic figures for the area: 78 percent are Hispanic and 11 percent are Asian.

Another indicator of the central location of immigrants is the population make-up of Los Angeles's traditional Black ghetto. This older residential area, built up by the 1920s, includes Avalon, Baldwin Hills, Central, Exposition Park, Green Meadows, Leimert Park, Santa Barbara, South Vermont, Watts, and West Adams (see Figure 3.2). The percentages of new and old immigrants in this south-central part of the city are almost equal to the county figures. With 21 percent of the population, Hispanics are by far the largest non-Black group in the traditional ghetto.

Between 1970 and 1980, the net Hispanic growth was 117 percent; during the same period the Black population declined by 8.3 percent (Oliver and Johnson, 1984). It is indeed rare for immigrants to have penetrated a high-density Black urban area to such a degree that they have begun to displace its residents. In sum, even the presence of low-income Blacks residing in the core of the city has not deterred many new immigrants from entering the areas adjacent to the downtown business district, just like immigrant waves before them.

Are the new immigrants more residentially segregated from natives than from older immigrants? The answer is yes. About 41 percent of all new immigrants would have to move in order to become fully integregated with the native population, whereas only 25.5 percent of the new immigrants would have to move to achieve an even distribution with old immigrants. (And surely the level of segregation between new immigrants and natives is higher than 41 percent, since the census counts native offspring of foreign-born parents as natives and this lowers the segregation index.)

INTERETHNIC SEGREGATION

The next questions focus on whether the residential patterns of Hispanics and Asians more closely resemble European ethnics or Black

TABLE 3.4   Indices of Dissimilarity for Ethnic and Racial Groups
in Los Angeles County: 1980

	Anglo	Hispanic	Black
L.A. County			
Hispanic	56.9		
Black	80.7	72.4	
Asian	49.2	53.7	83.5
L.A. City			
Hispanic	62.2		
Black	84.9	72.7	
Asian	51.7	44.2	78.1
Outside L.A. City			
Hispanic	53.3		
Black	76.9	72.0	
Asian	48.9	64.9	91.0

SOURCE: Garcia (1985).

migrants. Typically, European ethnics have not been highly segregated—
or, if highly segregated, then only for a short time. The segregation of
Blacks has been consistently high and unchanging (Lieberson, 1963).
And when Blacks move to the suburban ring of a city, these outlying
communities quickly tend to become high-density Black areas.

In Los Angeles County Blacks have also been more segregated from
Anglos than either Hispanics or Asians. Van Arsdol and Schuerman
(1971) reported that by 1960 it was evident that new and old Black
enclaves had already developed ghetto characteristics and that Anglos
had abandoned the inner city. They also indicated that the lower rates of
segregation of Hispanics and Asians were partly due to their location in
adjacent suburbs and satellite cities. Their findings from the 1940 to
1960 census data, however, did not exclude the possibility of increased
segregation of Hispanics and Asians, since this suburban dispersal had
not kept pace entirely with the overall rate of metropolitan expansion.

What is the current pattern of interethnic segregation? Is it stable
throughout the county? Our analyses suggest that the basic pattern of
segregation from Anglos in 1980 is consistent with the postwar pattern.
Blacks are the most isolated and, though to a lesser degree, Hispanics and
Asians experience significant levels of segregation. All three non-Anglo
groups are spatially isolated from Anglos at levels higher than those

TABLE 3.5 Population Movement Based on Replacement Ethnic
           and Racial Groups in Los Angeles County: 1980
           (in thousands)

	Anglo	Hispanic	Black
L.A. County			
Hispanic	1,563.1		
Black	1,231.9	929.7	
Asian	404.1	398.4	513.1
L.A. City			
Hispanic	645.4		
Black	629.6	452.8	
Asian	179.1	141.6	174.4
Out L.A. City			
Hispanic	898.9		
Black	495.1	456.2	
Asian	150.6	113.0	274.2

SOURCE: Garcia (1985).

experienced by European ethnics. Blacks are also highly segregated
from the other two non-Anglo groups. Interestingly, Hispanics and
Asians are no more segregated from each other than they are from
Anglos. These patterns of interethnic segregation are more or less the
same in the city and in the suburban areas outside the city. Inside the
city, however, Black and Hispanic isolation from Anglos appears
significantly higher than outside the city (i.e., by differences of more
than five percentage points). Among the non-Anglo groups, only Asians
appear to be more segregated in the suburban ring than in the central
city (see Table 3.4).

How many people would need to move in order to eliminate the
various forms of interethnic segregation found in the county? On the one
hand, about 1.5 million persons would have to move to achieve
complete Hispanic/Anglo integration, and 398,000 would have to move
to achieve complete Hispanic/Asian integration. And whether in city or
suburb, the elimination of Hispanic/Anglo segregation would require
the greatest movement of people. Within the city, the desegregation of
Hispanics or Blacks from Anglos seems to be an equally large order. The
least movement would be required in order to affect Asian/Anglo
segregation (see Table 3.5).

Because ethnic/racial identifiers have changed over the years, and researchers have not always used the same units of analysis, it is difficult to reconstruct precisely past decennial changes in interethnic segregation for Los Angeles. But overall, both Blacks and Hispanics appeared to be more segregated in 1970 than in 1960 period (compare Grebler et al., 1970; Massey, 1979a, 1979b; Rabinovitz, 1975; Van Arsdol and Schuerman, 1971; Van Valey et al., 1982). The next logical question is: Are the 1980 levels of ethnic/racial segregation higher than in 1970?

A study conducted by the Southern California Association of Governments indicates that the overall spatial isolation of Anglos remained constant between 1970 and 1980 and that Los Angeles County is the most segregated county division in all of Southern California (*Los Angeles Times*, 1982). Other findings are that Hispanics and Asians were more isolated in 1980 and that communities with a well-balanced ethnic/racial mix still are very rare. It appears that Black/Anglo segregation fell 10 percent and Hispanic/Anglo and Asian/Anglo segregation each rose about 10 percent.

Results from Akulicz de Santiago (1984) analyses of central cities indicate that Hispanic/Anglo segregation increased by 6 percent in the city, while both Hispanic/Black and Black/Anglo segregation declined by 10 percent. This suggests that Hispanic/Anglo segregation increased more in the suburbs than in the city.

ETHNIC/RACIAL INVASION AND SUCCESSION

In many cities, non-Anglos are increasing more rapidly than Anglos. In such cities the most usual outcome is that new minority residents enter (or invade) and take over tracts where they were previously underrepresented while holding onto tracts in which they were overrepresented. This is the classic pattern of immigrant invasion and succession. The rise in some of the segregation indices implies that some ethnic/racial residential succession took place in the last decade. The question now is: What was the overall pattern of invasion and succession?

Massey (1983) compared 1960 and 1970 Los Angeles data for Anglos, Blacks, and Hispanics. He found some significant differences in ecological patterns for Blacks and Hispanics: Invasion followed the entry of Hispanics into new neighborhoods, but Hispanic succession was not inevitable or even likely in many residential areas. An implicit conclusion was that Anglo-Hispanic segregation is muted by the absence of inmigrant succession.

TABLE 3.6   Racial and Ethnic Composition of Census Tract
Clusters for the City of Los Angeles: 1950-1980
(in percentages)

Cluster	Tracts[a]	Anglo	Hispanic	Black	Asian[b]	HI
Anglo						
1980	44.2	78.8	15.2	3.4	2.6	35.4
1950	88.0	92.2	4.5	2.3	1.0	14.2
change	−43.2	−13.4	10.7	1.1	1.6	20.2
Hispanic						
1980	22.0	18.2	67.4	6.0	8.4	50.2
1950	3.5	35.4	52.8	6.7	5.1	58.9
change	18.5	−17.2	14.6	−0.7	3.3	−8.7
Black						
1980	18.3	1.3	7.1	89.2	2.4	19.8
1950	4.5	10.6	8.8	77.9	2.7	37.3
change	13.8	−9.3	−1.7	12.2	−0.3	−17.5
Asian						
1980	5.0	12.8	38.5	11.3	37.4	68.2
1950	0.1	54.5	7.3	6.7	31.5	59.4
change	4.9	−31.7	31.2	4.6	5.9	8.8
Mixed						
1980	9.9	47.2	27.2	12.7	9.9	68.7
1950	3.8	46.3	19.1	20.8	13.8	67.7
change	6.1	0.9	8.1	−8.1	−0.9	−1.0

SOURCE: Community Analysis Bureau (1977).
a. The numbers of tracts analyzed for 1950 and 1980 were 710 and 695, respectively.
b. Contains a small number of American Indian and other enumerations.

The figures in Table 3.6 show several aspects of ethnic/racial invasion
and succession in the city of Los Angeles. This table shows tracts
clustered by the largest ethnic/racial group present. A "mixed" cluster is
tracts where no one group dominated (i.e., composed 50 percent or more
of the population).

The change figures in the column labeled "tracts" summarize the
succession that occurred. Between 1950 and 1980, the number of Anglo
tracts fell by 43.2 percent. We note that 18.5 percent became Hispanics;
13.5 percent Black; 4.9 percent became Asian, and 6.1 percent mixed
(with the largest non-Anglo group being Hispanics). This turnover was
almost entirely confined to the most central part of the city (Community
Analysis Bureau, 1977).

In most cases, the composition and degree of overall diversity within each ethnic/racial cluster changed noticeably. Anglo neighborhoods are now more heterogeneous. In 1950, the likelihood that two neighbors in Anglo tracts were of different groups was just 14.2 percent; in 1980 the likelihood is about 20.2 percent. The increased presence of Hispanics is the main reason for the greater diversity. Despite a large loss of Anglo residents, Asian clusters also increased in their ethnic variety. Again, Hispanics accounted for the greater diversity. On the other hand, Hispanic and Black clusters are now less heterogeneous than they were in 1950.

All of the non-Anglo groups broadened their geographical dispersion between 1970 and 1980. Asians might have made the greatest number of new entries into the suburbs of the groups (*Los Angeles Times,* 1982). But in most cases new Asian residents did not exceed 5 percent of the indigenous population in areas where they were initially few in numbers. For the suburbs, Asian invasion really only took place in the Hispanic area of Monterey Park. In 1980 they accounted for 33.6 percent of the population of this area, a twofold increase from 1970. Black succession appeared to have taken place in the south-central suburbs of Carson and Gardena, but there also has been a steady growth of Asians and Hispanics in these communities. On the whole, Black and Asian invasion and succession into suburban areas has been very limited between 1970 and 1980.

In the last decade, the Hispanic population continued its expansion outside the city. Table 3.7 displays the Hispanic concentrations in selected suburbs where they have had relatively high visibility for the past twenty years. All 21 of these communities are located on the east or southeast side of the city. In 1960, only in East Los Angeles, the city's traditional barrio, were Hispanics the numerical majority. By 1980, ten additional communities had Hispanic percentages in excess of 50 percent, and at least four more communities had 50 percent concentrations. In contrast to Hispanic growth, other populations declined. Only in the communities of Hacienda Heights and Paramount did the Anglo population grow in actual size between 1970 and 1980. By most definitions these figures suggest a pattern of widespread suburban succession (see, for example, Taeuber and Taeuber, 1965).

SOCIOECONOMIC INTEGRATION AND ETHNIC DISPERSION

Our last ecological question concerns the relationship between socioeconomic status and ethnic/racial segregation. After exploring the

TABLE 3.7   Percentage of Hispanics in Los Angeles County for
Selected Areas Outside City Limits: 1960-1980

	1960	1970	1980	Non-Hispanic Change [a]
Alhambra	4.3	17.4	37.6	−
Azusa	17.5	35.3	42.3	−
Baldwin Park	8.7	30.0	58.1	−
Bell Gardens	2.2	19.7	64.3	−
Carson	7.8	18.3	23.3	+
City of Commerce	41.0	69.0	85.0	−
Compton	9.0	13.5	21.1	0
East Los Angeles	67.9	87.2	94.1	−
El Monte	13.7	31.4	61.4	−
Hacienda Heights	8.9	20.9	26.4	+
Huntington Park	4.5	35.9	80.7	−
La Puente	21.7	46.2	62.5	−
Lynwood	3.5	16.1	48.5	−
Montebello	18.4	44.2	59.3	−
Monterey Park	11.6	18.1	38.8	+
Norwalk	15.0	25.7	40.1	−
Paramount	7.8	17.9	46.2	+
Pico Rivera	26.4	59.2	76.1	−
Pomona	9.2	16.3	30.1	−
Rosemead	20.0	34.8	57.4	−
San Gabriel	15.3	25.5	38.2	−

SOURCE: Garcia (1985).
a. This refers to non-Hispanic population changes for 1970-1980: a zero indicates a
change of fewer than 100 persons; a minus sign indicates a decrease in the absolute
number; a plus sign indicates an increase in the absolute number.

impact of education and family income on segregation in areas of sizable
Hispanic and Black communities, Massey (1978, 1979a) concluded that
Hispanic segregation from Anglos was mainly due to their low
socioeconomic position, but that was not true for Blacks. The forecast,
then, for Hispanics was that as the group improved its overall
socioeconomic position, the level of spatial isolation would decline.

TABLE 3.8   Indices of Dissimilarity, Conditioned by Education
or Family Income, for Ethnic/Racial Groups in Los
Angeles County: 1980

| | | White vs. | |
	Hispanic	Black	Asian
Education			
elementary (0-8 years)	68.2	90.7	69.8
some high school (1-3 years)	64.6	86.9	69.5
high school (4 years)	63.3	84.4	68.2
some college (1-3 years)	64.1	81.7	66.1
college (4 or more years)	66.6	81.7	67.4
Family income			
0-$4,999	66.6	88.0	66.5
$5,000-$7,499	66.7	87.3	71.3
$7,500-$9,499	66.2	87.3	71.9
$10,000-$14,999	64.4	85.3	66.3
$15,000-$19,999	64.4	85.3	66.3
$20,000-$24,999	62.9	85.1	62.8
$25,000-$34,999	61.5	84.9	61.5
$35,000-$49,999	61.9	84.8	61.9
$50,000 or more	71.2	85.1	66.0

SOURCE: Garcia (1985).

In Table 3.8 the level of segregation of Hispanics, Blacks, and Asians
from Anglos is calculated, when socioeconomic level is controlled. How
segregated are these groups from Anglos when the ethnics share levels of
education and income? Following the procedures of Massey (1979a),
five educational levels for adults and nine for family income were used.
Once these data are examined carefully, it becomes apparent that
segregation from Anglos does not decrease when any single ethnic group
shares socioeconomic status with Anglos.

HOW ACCURATE WAS THE OUTLOOK FOR 1980?

The data confirm the classical human ecology tenets regarding immigrant settlement. However, the findings and the information from other
studies are at odds with some predictions made about the path of ethnic
segregation in Los Angeles County. Instead of a decline in intergroup
isolation, a general increase seems to have occurred; and ethnic succession appears to be the course of Hispanic residential patterns rather than
spatial assimilation.

Why has the degree of segregation of Anglos remained high? And why has Hispanic and Asian segregation increased? Precise answers to these questions require a fuller analysis, but the data do reveal some telling points.

First, the enormous size and restricted geographical distribution of the new immigrants directly contributed to the higher levels of segregation by raising their concentrations in traditional ports of entry. At best this is only a small part of the story. Factors contributing to the decline in the absolute numbers of Anglos, such as outmigration and low fertility, are equally noteworthy determinants of the increased ghettoization of the county.

The change-figures in Table 3.3 help to illustrate the importance of both in- and outmigration patterns. In the last decade, the average annual population change for Anglos was a loss of 66,000 people. The comparable change-figure for the combined Hispanic and Asian population was a gain of 83,000 people. Thus for every net gain of 10 Hispanics or Asians, there was a net loss of 8 Anglos.

In this light, it is easy to see why predictions about segregation in 1980 did not materialize. To the extent that forecasters were unable to foresee the divergent changes in the Anglo and non-Anglo populations, their predictions were destined to be off target.

Another factor that may also have altered the predicted path of ethnic/racial segregation was the decline in metropolitan growth. With a net population gain of 6.2 percent in the 1970s, Los Angeles County no longer appears to be an expanding locale. The surrounding counties of Orange, Riverside, San Bernardino, and Ventura, in comparison, had gains that ranged from 32 percent to 45 percent, and their accumulated gain accounted for 72 percent of the net gain for the five-county area. More so than ever, desegregation in Los Angeles must be a function of population redistribution rather than equal access to newly developed residential neighborhoods. If ethnic/racial segregation does not vary across socioeconomic statuses (Table 3.8), then the likelihood of any large-scale intergroup redistribution in the near future is most unlikely.

DISCUSSION

A central concern of many who have researched interethnic segregation is the extent to which minority-group isolation impedes economic progress and overall social integration. The new immigration will certainly resurrect old themes regarding intergroup conflict, cultural adaption, and individual and structural assimilation for future research programs. Although a direct response to any of these open-ended issues is beyond our present means, we would like to conclude by making some brief general comments about how this new era of invasion and succession of "fourth wave" immigrants might affect the direction of interethnic relations.

Community-based resistance to ethnic/racial invasion is often linked to the degree of solidarity found among the older residents (e.g., McKenzie, 1967). And as Logan and Stearns (1981) conclude, the most common motive cited for exclusionary practices is the fear of status deprivation. It is a fear among present occupants that their community will lose status if the homogeneity of the population is altered. At the heart of the matter then is how people perceive out-group members vis-à-vis residential change.

Certainly Hispanics and Asians share characteristics that have historically exacerbated social divisions in the United States. Both groups are visibly different and also culturally distinctive. For instance, in 1980, only 19.8 percent of Hispanics and 26.4 percent of Asians residing in the county were English monolingual speakers. And in socioeconomic terms, Hispanics do not do very well in terms of completed schooling and personal or family income, although Asians tend to approach the norm. Paradoxically, Hispanics continue to attain higher annual income levels than Blacks despite their much lower educational attainments (see Appendix Table 3.A4).

How are ethnic/racial groups ranked in terms of social distance? Table 3.9 displays the subjective ranking of ten ethnic/racial categories by a national probability sample of Black and Anglo respondents and a separate probability sample of Black and Anglo respondents residing in Southern California. Feelings toward a group could range from 0 to 100, with 100 being the most favorable response. In most cases the response patterns of each sample were identical.

With the exception of the American Indian category, all the non-Anglo groups were ranked below the European categories. Interestingly, the groups that denote immigrant status (i.e., Chinese through Vietnamese) rank below the categories that denote native populations. More interesting, the two groups that denote refugee status (i.e., Cubans and Vietnamese) were ranked the lowest of all. How much these rankings actually reflect anti-ethnic or anti-immigrant feelings is, of course, arguable. Yet the response pattern does parallel an immigrant-to-nonimmigrant social distance continuum.

TABLE 3.9   Feelings Toward Different Ethnic and Racial Groups

	Total	Southern California
Very favorable	100.0	100.0
English descent	75.3	76.6
Irish	72.7	74.4
American Indian	73.6	72.8
German American	70.5	70.8
Blacks	67.3	67.5
Chinese	63.5	66.1
Mexican American	63.5	65.9
Puerto Rican	56.5	58.1
Cuban	53.1	55.3
Vietnamese	56.1	55.1
Total	1378	96

SOURCE: Cardoza et al. (1984).
NOTE: Data are from a national probability sample of Anglo and Black respondents. Southern California respondents resided in Los Angeles and San Diego.

A now familiar concern in many urban centers is a negative backlash response by Anglos with regard to minority groups and social change. There is also a growing concern about possible tension between lower-status minorities and the latest wave of immigrants. For example, the argument has been raised that recent entrants have threatened the economic position of Blacks (e.g., Jackson, 1979).

Is there a minority-backlash attitude by Anglos in Los Angeles? And how do Blacks evaluate their economic status in contrast to immigrants groups? Our Southern California data indicate that 38 percent of Anglos in Los Angeles believed that minorities are given too much special consideration. A slightly higher percentage felt that Blacks have only a few problems in getting hired or promoted. About 14 percent thought that Hispanics enjoy better chances for getting ahead. In the national sample, 26 percent of the Black respondents thought that their chances for getting ahead were less than for Hispanics (see Table 3.10).

Observations from Oliver and Johnson's (1984) study of Los Angeles inner city residents reveal a similar pattern of intergroup tolerance.

TABLE 3.10   Attitudes Toward Ethnic and Racial Minorities, by Race (in percentages)

| | Total | | Southern California |
	Anglo	Black	Anglo
How much special consideration is being given to racial minorities right now?			
too much	42	09	38
right amount	33	21	40
too little	25	70	22
Would you say Blacks miss out on jobs or promotions because of discrimination?			
none	07	0	07
a few	36	0	36
some	30	33	30
many	27	67	27
Compared to Hispanics, do you think the chances for getting ahead for are much worse?			
much worse	03	07	05
worse	07	19	09
same	45	51	34
better	22	17	30
much better	23	06	22
Total	1217	168	92

SOURCE: Cardoza et al. (1984).
NOTE: Data are from a national probability sample of Anglo and Black respondents. Southern California respondents resided in Los Angeles and San Diego.

Focusing on the issue of undocumented immigration, 79 percent of Anglo and 66 percent of Black residents felt that undocumented Mexican workers took jobs away from native workers. And 59 percent and 35 percent of Anglos and Blacks thought Mexican Americans did not try hard enough to learn English.

It might very well be that there are minority-majority and inter-minority hostilities, with the former possibly being greater. What still must be determined is the degree to which the populace at large distinguishes between a variety of intragroup statuses: ethnic versus immigrant, new immigrant versus old immigrant, documented alien versus undocumented alien.

TABLE 3.11   Indexes of Dissimilarity for Los Angeles Unified
School District: Fall 1966 to Fall 1978

	Blacks	Anglo versus Hispanics	Asians	Blacks versus Hispanics	Asians	Hispanics versus Asians
1966	91	66	67	84	76	59
1967	91	65	67	85	76	59
1968	91	65	65	85	77	57
1969	90	64	65	85	78	58
1970	89	64	65	85	78	57
1971	89	64	65	85	78	57
1972	87	64	60	83	76	56
1973	86	64	61	82	76	56
1974	84	65	60	80	75	56
1975	83	65	58	80	76	56
1976	81	66	57	78	75	55
1977	77	66	55	76	73	56
1978	71	62	50	73	70	54
Average Change 1966-1977	−1.3	0.0	−1.1	−0.7	−0.2	−0.3
Change 1977-1978:	−5.8	−4.0	−5.0	−3.0	−3.1	−1.7

SOURCE: Farley (1979).

Yet another way to approach intergroup isolation is to examine public school segregation. Since 1954 the tolerance of Anglos for racial desegregation of public schools has served as a useful barometer in evaluating mainstream views about non-Anglo groups. There are published data on the level of isolation experienced by minority students attending public schools in the city. These data are particularly informative because they span a period when the school district implemented a variety of desegregation programs. Therefore we can ask: To what extent have minority students become more or less integrated than the population as a whole?

In the Los Angeles Unified School District, segregation can still be found throughout the system. In 1977, for example, fully 62 percent of minority students in elementary attended schools that had 90 percent to 100 percent minority enrollment; for junior and senior high schools the comparable figure was 51 percent (Rabinovitz, 1978). As with the residential population, student segregation also differed by ethnic/racial group: 25 percent of Asians, 54 percent of Hispanics, and 82 percent of Black students attended schools where the Anglo enrollment was less than 10 percent (Rabinovitz, 1978).

Between 1966 and 1977, the district's desegregation efforts were entirely voluntary. During this time there was a decline in non-Anglo segregation, but the pace was slow. The average yearly decline in non-Anglo/Anglo segregation was a scant 1.3 percent for Blacks, 1.1 percent for Asians, and less than 1 percent for Hispanics. One year after the 1978 inauguration of court-mandated desegregation policies, the average non-Anglo/Anglo segregation level declined by an additional 5 percent. For the observed period, there were small gains made in non-Anglo/Anglo student desegregation. Yet despite the implementation of numerous integration programs over a thirteen-year period, Hispanic and Asian students appeared just as segregated as their nonstudent counterparts (compare Tables 3.4, 3.8, and 3.11). The 1978 ID score for Black students is about 10 percent lower than the 1980 ID score for the entire Black population. Thus only the most segregated student population made gains in desegregation.

What was the response of Anglos to increasing minority enrollments (especially of immigrants) and greater pressures to integrate? One response was to leave the school district.

Prior to the involuntary integration policies adopted by the district the average ethnic/racial change in annual enrollment was –5.7 percent for Anglos, 6.7 percent for Hispanics, 0.6 percent for Blacks, and 4.3 percent for Asians. During the first year of court-imposed integration policies, the non-Anglo groups experienced changes that were, more or less, in line with their rates recorded for the previous year. However, Anglo student enrollment declined by 15 percent.

Taking into account the differential ethnic/racial growth rates in enrollment, Farley (1979) estimated the magnitude of "Anglo flight" that may have been induced by the inauguration of court-ordered integration plans. He concluded that during the 1977 academic year, as many as 5 percent of the Anglo student population fled the public school system in anticipation of expanding integration policies. And an additional 7 percent of the potential 1978 Anglo enrollment left before the first year of the mandated desegregation program. The drop in Anglo enrollment for schools actively participating in desegregation activities was 30 percent. For nonparticipating schools the comparable change rate was 8 percent.

Ultimately, the majority response to school desegregation was effective resistance. By 1982, the mandatory policies had been legislatively dismantled.

The school desegregation case in Los Angeles exemplifies the importance of our summary comments. First, there are multiple forces at work that help to make ethnic/racial segregation the status quo in U.S. metropolitan areas, and these forces include both ecological and nonecological factors; that is, certain population populations changes can be expected as a neighborhood ages, but preferences and organized opposition to perceived changes can promote or deter the process of invasion or succession by new out-group members. Second, research on urban ecology must adopt a flexible multiethnic perspective. As Crain (1978) states: "Desegregation in Los Angeles is made more complex by the presence of native and foreign-born Mexican Americans, immigrants from other Hispanic countries, American Indians and Asian Americans, indigenous and from abroad." Future research on the centralization of immigrant communities and the reproduction of ethnic/racial segregation in the suburbs should consider these points.

(Appendix tables begin on page 98)

TABLE 3.A1 Description of Demographic Measures

Index	Equality	Description
Index of Dissimilarity	$ID = \frac{1}{2} \Sigma /(H_i/H) - (B_i/B)/$	where H_i and B_i are the population subtotals of two ethnic categories in the i-th areal unit, and H and B are the population total of each group (i.e., $H = \Sigma H_i$ and $B = \Sigma B_i$). The result is multiplied by 100. The ID can be interpreted as the percentage of either group that would have to move in order to elominate segregation between the groups and therefore produce an ID score of zero.
Replacement Index	$RI = 2q(1-q) \, ID$	where q equals the proportion of the numerically smaller group (i.e., if $B < H$, the $q = B_i/B$). The result is multiplied by 100. The RI can ne interpreted as the percentage of the total population (H + B) that would have to move to produce an ID equal to zero if both groups relocated to areal units where they were under represented.
Heterogeneity Index	$HI = 1 - \Sigma(H_i/T) (H_i-1/T)$	where H_i is a vector containing each ethnic subtotal and T is the total population figure for a specified areal unit. The result is multiplied by 100. The HI can be interpreted as the likelihood that two residents selected at random are not of the same ethnic category.

TABLE 3.A2 Distributions of the Foreign-Born Population, by Year of Immigration for Selected U.S. Counties: 1980 (in percentages)

Year of Immigration	Los Angeles	Orange	San Francisco	New York[a]
1975-1980	34.8	36.3	30.5	19.7
1970-1974	22.2	17.1	16.2	17.8
1965-1969	13.0	11.7	12.9	15.5
1960-1964	8.8	10.7	9.1	9.0
1950-1959	9.6	12.8	12.3	11.7
Before 1950	11.6	11.5	18.9	26.3
Total percentage	100.0	100.0	100.0	100.0
Total number	1,664,800	257,200	192,200	2,046,400
Percentage foreign-born	22.6	13.3	28.3	19.6

SOURCE: U.S. Bureau of the Census (1983a, 1983b).
a. The New York part of the New York-New Jersey Standard Metropolitan Statistical Area.

TABLE 3.A3 Percentage Ethnic and Racial Distributions for Selected U.S. Counties: 1980 (numbers in thousands)

	Los Angeles	Orange	San Francisco	New York[a]
Anglo	53.3	78.9	53.2	64.0
Hispanic	27.0	14.5	11.2	13.6
Black	12.6	1.3	12.7	19.3
Asian	6.1	4.9	22.0	2.7
American Indian	0.7	0.2	0.5	0.2
Other	0.3	0.2	0.4	0.2
Total percentage	100.0	100.0	100.0	100.0
Total number	7,477.5	1,919.5	678.9	10,673.7
Heterogeneity Index	63.3	35.3	64.1	53.4

SOURCE: U.S. Bureau of the Census (1983a, 1983b).
a. The New York part of the New York-New Jersey Standard Metropolitan Statistical Area.

TABLE 3.A4 Educational and Income Attainments of Ethnic and Racial Groups in Los Angeles County: 1969-1980

Ethnic	Median Educational Attainment		Median Family Income (dollars)	
	1970	1980	1969	1979
Anglo	12.4	12.4	11,791	23,217
Hispanic	9.3	9.7	8,466	15,531
Black	12.0	12.1	7,571	14,879
Asian[a]	12.8	12.8	13,768	23,676
American Indian	11.8	12.0	8,342	17,890

SOURCE: Crain (1977), U.S. Bureau of the Census (1983c).
a. 1970 data reflect Chinese and Japanese only.

4

Improving the Data:
A Research Strategy
for New Immigrants

JOSE HERNANDEZ

☐ CONTEMPORARY PARADIGMS of urban composition and change in the United States rely primarily on population analysis for quantitative evidence. However, efforts to account for recent variations resulting from the new immigration are difficult to carry out. Serious problems of validity and reliability come from the incomplete and defective condition of demographic information. A strategy is here proposed for improving population data, with particular emphasis on the micro-level of analysis so crucial to theoretical revisions. The current and anticipated needs of new immigrants for population data of use in their participation in public policy are also considered.

THE QUESTION OF NUMBERS

Major methodological issues begin with the most fundamental matter of determining how many new immigrants live in the nation's cities. Past and proposed changes in the scope and definition of census data on race, ethnicity, foreign birth, parentage, and ancestry debilitate or invalidate designs based on longitudinal comparisons. Delays and curtailment of report publication and tape availability have further lessened the value of the 1980 enumeration as a measure of a rapidly changing subject. Most recent Census Bureau publications on ethnicity

AUTHOR'S NOTE: *I wish to express my gratitude for the competence and helpfulness of the research staff at the Latino Institute.*

pertain to native minorities or are limited to nationalities typical of the old immigration (for example, see U.S. Bureau of the Census, 1973). Except in 1979, the Current Population Survey has yet to address immigration in its ongoing data production agenda. Little or no follow-up is apparent in research on the role of immigration in the undercount of the nation's population, a topic of considerable interest before the 1980 census. Such a development depended partly on the improvement of statistical sources that have largely retained the deficiencies identified years ago. Vital register data essential to the study of the demographic consequences of the new immigration remain nonexistent, fragmentary, or subject to error. Little or no information is available on the urban settlement patterns of new immigrants, their continued geographic movement, and return migration. Data from visa applications provide some indication of intended place of residence, and some figures are available from the annual registers of aliens. But these sources have serious validity problems in determining the extent and nature of urban change deriving from the new immigration, and cannot be effectively related to census results.

A second and enormous gap in our knowledge is the continuing lack of a valid and reliable measure of unrecorded foreigners in the national and local populations (U.S. Bureau of the Census, 1982b; U.S. Library of Congress, 1980: 3-20). Without much evidence, most sources equate the unrecorded with the undocumented. However, estimates of the undocumented are inconsistent and cannot be validly related to census data, which, of course, are gathered without regard to the legality of residence. Aside from their alleged but unproven evasion of official data collection, our hearsay knowledge about the undocumented suggests that those counted represent only a fraction of the many who enter and leave the nation or a particular city during a given period of time (U.S. Select Commission on Immigration and Refugee Policy, 1981: 477-490; hereafter cited as USSCIRP). Few or none of the variables determining undocumented migration have been measured or studied in preparation for data collection. These certainly should include the following:

- the demand for and difficulty of entry, both actual and perceptual
- such happenings as now attributed to "chance": evasion of detection, sponsoring relatives, recruiters from the United States, and so forth
- receiving patterns in urban American society, particularly the labor market and socioeconomic relations with legally admitted immigrants and the native population

A third information gap necessitates the development of reliable return migration rates. Based on very tenuous evidence, a 30 percent return rate is generally cited (USSCIRP, 1981: 283). But this and similar data remain questionable, considering the circulatory movement between the United States and nearby nations (Bonilla, 1984). Communication and transportation facilities have greatly simplified continued ties and involvement with these places of origin. We are just beginning to gain knowledge on such related topics as network migration, return economic ventures, and the significance of remittances from the United States. To obtain a verifiable return migration rate (especially for particular nationalities), this knowledge must be considerably expanded. Since no information is collected at departure from the United States and the return of the undocumented may be as clandestine as their entry, measurement must be indirect and innovative in methodology.

The fourth gap—data on local geographic areas—poses an even greater obstacle to the revision of paradigms of urban composition and change. In the city of Chicago, for example, the 1980 census counted 422,063 persons of Spanish origin, a figure updated to 526,000 in 1983 by conventional demographic methods that considered the undercount and estimates of population change, including net migration (Latino Institute, 1983b). In this project the available demographic information suggested that the population measured was composed largely of U.S. natives and legal residents. Moreover, the U.S. Immigration and Naturalization Service has been said to estimate up to 450,000 undocumented Hispanics in Chicago, adding that the INS "can go out on any given day and pick up as many illegal aliens as we have time to put in vans" (Kiefer, 1984: 133). In the past, scholars have considered such estimates to be exaggerated guesswork. Practical research experience suggests, however, that they represent social realities that must be somehow reconciled with data available from traditional sources to obtain a true picture of Chicago's Latino population.

If informed and intuitive estimates are considered complementary (instead of contradictory) evidence, we obtain a notion of magnitude corresponding to the sense of community gained by actual contact with new immigrants. Inasmuch as 200 Latino service organizations exist in Chicago, the problems and demands of the population seem greater than one would expect of a half million people. If census figures were correct, the profusion of grocery stores and restaurants catering to new immigrants by nationality and location of origin would be much smaller. Such expressions of presence offer a basis for the figures proposed by sources generally considered "unscientific." Their value is not so much a corrective for the undercount, as evidence of population realities not yet measured by official data collection procedures.

As further illustration, 119,540 Dominicans and 42,980 Colombians were counted by the 1980 census in New York City (Mann and Salvo, 1984). Nondemographic sources in 1978 had placed the Dominican population in the 300,000 to 400,000 range, observing "phenomenal" growth by direct migration and the fifteen-minute air trip to U.S. territory in Puerto Rico. The Colombian expressions of presence (soccer clubs, housing rehabilitation, civic organizations) clearly indicate that estimates greater than 200,000 are reasonable (Ugalde et al., 1979: 236; Sassen-Koob, 1979: 316). Some researchers hesitate to look beyond conventional data sources partly because of the special meanings that population numbers have for certain nationality groups. A half million Dominicans represent a tenth of the entire Dominican population—with vastly different demographic implications than, say, a half million immigrants from a giant population such as that of India. Colombian immigrants in the United States may be less numerous than those in Venezuela, but in both instances the issue of "how many" is tied to the timely and accurate measure of outmigration from Colombia. Considering the frequency of circular migration in some of the new immigrant populations, an innovative demographic methodology is needed to measure the consequences of the population transfer, for both totals and characteristics in the United States and in the nation of origin.

Then again, new immigrant groups have not been generally affected by the emphasis on accurate population figures that came with revenue sharing. Other more recent and urgent concerns are exemplified by the 20,000 annual per nation visa ceiling for legal entry to the United States, which places a constraint on the flow of immigrants from the principal nations of origin and is often said to be a factor in the movement of undocumented immigrants. Certainly the intricacies of immigration law and its application claim considerable attention by the people involved in the migration (Garrison and Weiss, 1979). Moreover, the first-come, first-served directive has resulted in a multitude of relatively small immigrant groups, each with a different sense of ethnicity and relation to public policy in the United States, including the issue of "how many." Recent events and impending changes in American public policy have drastically altered and will likely change the volume and composition of certain nationality groups. For example, anyone familiar with the

Haitians will agree that a national census total of 90,223 in 1980 missed the real total by a wide margin (U.S. Bureau of the Census, 1983: 8). In 1984, estimates ranged from 300,000 to 500,000 Haitians in the New York metropolitan area alone (Howe, 1984). As is so often found, the issue of numbers is further related to social divisions internal to the Haitian community. In addition to the standard demographic categories of first and second generation, Haitians identify themselves according to language (Creole or French), resident status (immigrant or refugee), and the usual legal and undocumented split (Buchanan, 1979: 298-313). But regardless of such distinctions, the generally negative publicity given to the Haitians merits much greater community attention than the question of numbers.

From these considerations we must conclude that the elementary question of "how many?" cannot be adequately addressed in traditional demographic analysis at present. The first corrective must be a comprehensive and effective reform in national data sources. In addition to improvements in the deficiencies noted, questions on immigration should figure prominently in preparations for the 1990 and 2000 censuses. However, allowing for the usual delivery schedule, data needed for a national assessment would not be initially available from this source until 1992 or later. The lack of comparable or independent measures poses an additional limitation on the use of the 1990 census for projections and similar demographic research methods. Experience with native minorities shows that a coordinated improvement in vital register data and repeated study in national surveys are needed to approach adequacy of enumeration, at least to the extent of distinguishing error from genuine variations in the data produced. This implies that the time of full enablement for research on the new immigration will be during the years after the 2000 census.

A FOCUSED APPROACH

Except as data on new immigrants might be available by urban geographic categories in the 1990 census, no detailed information can be currently expected from traditional sources for the revision of paradigms. A more proximate and supplementary research plan is therefore necessary, centering on a localized data collection effort to produce information by community area, census tracts, and, perhaps, even blocks. The beginnings of such research can be observed in the endeavors of new immigrant groups to complete neighborhood population studies and in the New York City Department of Planning survey of new immigrants, to be completed in 1985. The following considerations were derived mainly from the author's experiences in organizing the collection of longitudinal microdata for Latinos in Chicago (Latino Institute, 1984). The author supplemented this by observing and conducting purposive interviews with new immigrants in the city, suburbs, and satellite cities of the New York metropolitan area in 1984.

We must begin with the assumption based on accumulating evidence that (except for farm laborers and certain refugees) new immigrants typically originate in the largest cities of their nations of origin (USSCIRP, 1981: 282). For example, a recent study conducted in New York City found that 90 percent of the Colombian and 76 percent of the Dominican immigrants interviewed had grown up in urban areas, at least half in the four largest cities of their nations of origin (Gurak, 1983b: 17). Practically the same results were obtained from return immigrants to the Dominican Republic (Ugalde et al., 1979: 240). The literature on new immigrants makes it very reasonable to suppose that nationality groups from more distant places are at least as urban, or primarily metropolitan in background. They apparently arrive already familiar with the increasingly standard institutions of the world's largest cities: rapid transportation, supermarkets, newspaper advertisements, and the like. Many have the "urban frontier" orientation toward such cities as New York and Los Angeles: a contextual resemblance between the receiving city's image and the immigrant's ambitions for settlement and success in the United States (Bryce-Laporte, 1979: 214). The metropolitan connection is further strengthened by American mass media, international business and tourism, and both nationality and primary-group networks that facilitate immigration. All of these focus on a handful of large cities (New York, Los Angeles, Miami, San Francisco, Chicago, Houston, and Honolulu) and functionally related or nearby metropolitan areas. Incidental to this migration, some immigrants settle in major transportation hubs such as Denver, or large cities of particular significance to a nationality group. But these and other settlements tend to be of minor importance compared with the areas mentioned.

The metropolitan connection provides a regional delimitation of use in the implementation of short-range research strategies. To be more specific, records from the Immigration and Naturalization Service show that nearly half of the new immigrants admitted from 1975 to 1979 gave

TABLE 4.1   Population Totals and Distribution in States of Highest
Concentration, for Largest Nationality Groups Typical
of New Immigrants, 1980

Group	U.S. Total	Percentage Distribution by State					
		First		Second		Third	
Mexican	7,692,619	CA	44	TX	32	IL	5
Chinese[a]	910,843	CA	43	NY	16	HA	10
Cuban	803,226	FA	59	NJ	10	NY	10
Filipino	795,255	CA	44	HA	18	IL	6
Korean	376,676	CA	27	NY	9	IL	6
Asian Indian	311,953	NY	18	CA	16	IL	10
Jamaican	253,268	NY	54	FA	10	NJ	6
Vietnamese	215,184	CA	32	TX	11	LA	5
Dominican	170,698	NY	78	NJ	9	FA	4
Colombian	156,276	NY	34	FA	17	NJ	14

SOURCE: U.S. Bureau of the Census (1980).
a. Includes persons reporting Taiwanese ancestry.

New York and California as states of intended residence, with an
additional 30 percent specifying Florida, Hawaii, Illinois, New Jersey,
and Texas (USSCIRP, 1981: 235). Alien registration data for 1978 show
a somewhat lower proportion, but still over 60 percent reported
addresses in these seven states. Table 4.1 gives further evidence of the
consistency of geographic concentration among the largest new immigrant nationality groups. These data are based on the 1980 census
ethnicity or "ancestry" classification, which takes no account of birth-
place or citizenship, and thus measures the eventual patterns of settlement to a greater extent than the other statistics. Nevertheless, a very
similar pattern is found.

Table 4.1 also shows that the largest new immigrant groups are all
non-European and originate mainly in the Carribean, Mexico, and
Central America, or in the nations peripheral to the People's Republic
of China. This further serves to delimit the focus of short-term research
strategies to certain world-region and nationality group clusters. With a
minor exception for the Vietnamese, at least one major large-city
community is readily found in each cell that combines a state with a

specific ethnic category. It seems reasonable to assume that whatever changes are needed in paradigms of urban composition and change will be most apparent in these primary concentrations. Certain nationalities not listed in Table 4.1 tend to follow similar patterns. In Chicago, for example, the settlement of Central American immigrants is closely related to the residential distribution of Mexican immigrants. The neighborhood housing patterns of Thai immigrants resemble those of Koreans, and immigrants with advanced education tend to live in city or suburban settings similar to those of Filipinos and East Asians. In New York City, many Haitians live in areas with other Caribbean people of modest income, while West Indians who are occupationally mobile follow the residential patterns described by Bryce-Laporte (1979: 214) in regard to the long-established segments of the Jamaican community.

FIVE COMMUNITY TYPES

More specific direction to short-range strategies for demographic research on new immigrants is provided here. Since an ultimate objective is to facilitate revisions in paradigms of urban composition and change, it is crucial to link the state and metropolitan references just made with the local ecology of places where new immigrants have settled. The following suggestions are based primarily on recent literature and practical research experience in Chicago and New York. Five ecological situations are presented as a provisional typology of the residential concentrations most commonly found. This approach is exploratory, descriptive, and open to correction and further refinement, in step with the accumulation of future research results.

Before the changes in legislation that encouraged a large-scale entry of Third World immigrants, non-European foreigners settled mainly in neighborhoods segregated either to themselves or to non-European natives. Thus, a variety of Caribbean nationality groups have lived in Harlem, a community area of New York City well known for its American Black population (Foner, 1979; Lopez, 1974). From this experience emerged the concept of ghetto or barrio: a relatively stable, socially isolated ethnic and racial enclave, likened to the historic immigrant neighborhoods of White, European origin. The literature on the topic and research experience make it clear that this kind of immigrant residential pattern continues today. Nevertheless, the enormous volume and wide diversity characteristic of the new immigration have brought about significant changes in metropolitan patterns of settlement.

The first community type is still the ghetto or barrio. These areas have changed, however, as a result of the coexistence of different ethnic and racial groups in neighborhoods marked by the predominance of a single nationality. An example is Manhattan's Upper West Side (north of Harlem) where Dominicans are now the most likely nationality group found in a given street segment or neighborhood section. But various other immigrant and native groups are also present, either in highly localized clusters or dispersed throughout the area.

A variation of the first type is found where several nationality groups with cultural affinities live together in an area socially ascribed to them in general. About one-third of Chicago's central-city Hispanics live in the adjoining areas of West Town, Logan Square, and Humboldt Park. The original Puerto Rican barrios have blended with the settlement of new immigrants from Mexico and Central America, who now make up about 45 percent of the Hispanic population (Latino Institute, 1983a: 24-26). A street segment inhabited by Hondurans may be typically found along with a cluster of people from a certain locality in Mexico. Elsewhere, families of differing Hispanic backgrounds may live in the same apartment building. Grocery stores, laundromats, and churches serve as focal points for individual ethnic and geocultural groups. Although the area's external identity remains Puerto Rican, a Latino identity provides a basis for community interaction and a functional equivalent of nationality for dealing with outsiders who may be unfamiliar with variations in Hispanic backgrounds (Padilla, 1985).

Whether a single nationality or a cluster of nationalities predominates, the ghetto or barrio is limited in its capacity to absorb so many new immigrants. The other four community types have emerged from the settlement of new immigrants in neighborhoods where they replace the native population of long-term residents in the succession process. Given the general character of housing demands in large metropolitan areas, succession seems likely in neighborhoods no longer attractive to native residents—and not yet attractive to other groups that are more appealing to the real estate industry.

In New York and Chicago neighborhoods currently segregated to native non-Europeans (Blacks, Chicanos, and Puerto Ricans), there tends to be very little room for new immigrants. Housing scarcities stemming from abandonment and demolition of dwelling units and discrimination in other places have created barriers to settlement by outsiders. This means that new immigrants must seek housing in succession to native Whites of European origin, perhaps in competition with the native minorities. The extent to which new immigrants are preferred to native minorities when the outgoing population is White remains a major unanswered question. The current tendency of native minorities to seek housing beyond the central-city limits suggests that certain new immigrant nationality groups are preferred in neighborhood succession to long-term White residents of European origin.

In any case, it appears that the new immigrants are at the cutting edge of White flight from central-city areas. As they increase in number and are identified as a type of applicant, succession takes place as a function of family, friendship, and business ties that can be called "networks of neighborhood accommodation." The networks tend to blend with or replace the preferential system formerly operative in allocating dwelling and commercial units to Whites of a specific European nationality, religion, or social class. Since failures are common, new immigrants faced with hostility tend to seek out others of the same group, adding to the press for the expansion of a network in a receptive neighborhood.

The very large number of new immigrant nationalities means that some groups do not have networks large enough to predominate in such successions. In response to housing scarcities and discrimination, many cluster together in multiethnic succession. The results are visible in north central Chicago and the Brooklyn borough of New York, clearly showing the emergence of a *second community type: the heterogeneous new immigrant districts in which no single nationality group pre-dominates*. Generally speaking, this type of settlement has a more cosmopolitan ambiance than the ghetto or barrio and a greater potential for growth in the new immigrant population. It is marked by a variety of languages and expressions of presence in the form of stores, churches, organizations, and entertainment facilities.

The ecology of Haitian settlement in New York serves to illustrate differences between the first and second community types. As a nationality group subject to considerable discrimination, Haitians face extraordinary difficulties in finding housing and settling in central-city neighborhoods. In addition to their differences from European Whites in language, culture, and social class, Haitians differ from American Blacks and English-speaking West Indians as well. Many are not eligible for public housing, which makes up a major portion of the dwelling units in the older, native Black neighborhoods. High rents place private housing beyond the reach of many Haitians who might otherwise live in native Black ghettos.

As the population of Haitians has grown larger, openings for housing have occurred in the predominantly White ethnic neighborhoods in the Flatbush area of Brooklyn. These neighborhoods were on the fringes of the native Black ghettos—places where Whites perceived the likelihood of Black succession and were motivated to leave for this reason. The prices became moderate and the succession neighborhoods were attractive to Haitians as well. They consisted mainly of family homes with relatively spacious interiors and a backyard and basement. For an immigrant group with comparatively large households, an involvement in home industries, and a need to subdivide dwelling units, the brownstones and rowhouses offered advantages not generally found in the apartment buildings typical of native Black ghettos. The largest and most densely inhabited Haitian community is now in the Crown Heights/East Flatbush area. Here, as in the first type of settlement, Haitians tend to predominate in a given street segment or neighborhood section. Elsewhere in Brooklyn, however, Haitians share their neighborhoods with various other ethnic groups of new immigrants, following the second type of community pattern.

In the second community type of multiethnic and multiracial neighborhoods, the exterior appearance of buildings, shopping, and service functions all tend to reflect the new immigrant character of the population. A small number of natives live interspersed among the various nationalities—the Whites who remained, and Blacks and Hispanics from other areas of New York City. Given the strength of the new immigrant expressions of presence, further change seems unlikely in the near future. However, ethnic and racial coexistence of this nature is quite unusual in the urban ecology of large American cities. In time, perhaps a single nationality will predominate, as in the first community type. Or, instead, such areas may remain relatively heterogeneous and function as buffer or insulation zones between districts of predominantly native residents—those of European origin on one side and the minorities on the other. In certain places, they already fulfill this function.

Another possible future for the multiethnic, multiracial neighborhoods would be transition to the *third community type, which is distinguished by its instability and eventual succession by White middle-class settlement in the gentrification process*. The third community type also originates in new immigrant settlement of neighborhoods traditionally inhabited by White ethnic families of lower-middle and working-class standing. Succession takes place in the typical White flight pattern of aging and upward mobility; but it is essentially temporary in nature. The new immigrants are typically renters and hold little or no lasting tenure in neighborhood businesses. Such neighborhoods quickly become attractive to real estate speculators and eventually to White mainstream settlers who work in downtown locations and are often described as "young urban professionals," or "yuppies." In the gentrification process, neighborhoods initially deteriorate in physical appearance and housing facilities. However, as the turnover takes place, residential units are renovated in step with the replacement of new immigrant businesses by stores and services oriented to the incoming population of White middle-class Americans.

Perhaps the most visible example of the third community type is the strip of new immigrant settlement running northward from Old Town in Chicago to the adjacent satellite city of Evanston, Illinois. Various new immigrant groups have settled west of Lake Michigan in a mix of dense concentrations and heterogeneous configurations of race, ethnicity, and nationality. Observation and interviews led to the general conclusion that gentrification was moving northward at a minimum annual rate of two or three blocks, and westward, in certain places. As the turnover takes place, new immigrants are continually rolled over and drastic changes occur in the social and economic environment of the affected neighborhoods. Newly arriving immigrants and the people displaced by gentrification generally find housing at the fringes of the multiethnic strip, in neighborhoods typically inhabited by the long-term resident White ethnic families who tend to move out. Certain new immigrants are able to survive the gentrification process and remain living interspersed among the predominantly White middle-class Americans who move in. Incidental to the process, some native minority people of middle-class standing move in as well. The resulting ambiance is predominantly White, contemporary, and sophisticated—with variations typical of the downtown bureaucratic work force.

An important element of the gentrification process in Chicago is the nearby presence or coexistence with new immigrants of a substantial and predominantly White gay male community. The gays can be considered for our present purposes as the functional equivalent of an ethnic group. Faced with considerable discrimination in housing, the central-city workers of this community are segregated to the same kind of high-turnover neighborhoods as new immigrants. In certain very limited instances "gayification" takes place as the more affluent members of the community succeed in gaining ownership of businesses and residential units. But the typical situation resembles the rollover experienced by new immigrants. The gay expressions of presence in stores, churches, services, and places of entertainment continually move in step with the gentrification process. Thus a predominantly White population is subject to the same turnover in tandem by White successors whose lifestyle preferences are more socially acceptable to the American majority.

A fourth type of new immigrant community appears in the older neighborhoods of certain satellite cities of the metropolitan areas previously identified as the focus of short-range research strategies. Because conventional demographic data often distinguish between the central-city and a residual "other" category, this community type is easily mistaken for suburban settlement. In such cities as Aurora, Waukegan, and Joliet, Illinois; and Hammond, East Chicago, and Gary, Indiana, new immigrants live within the greater Chicago metropolitan area. Although the succession process may resemble the turnover in central-city areas, housing and social life patterns are different. These communities are also clearly distinguishable from the human environment of surrounding suburban municipalities.

As an illustration, from 1970 to 1980 the population of New York City fell by more than 10 percent, and the metropolitan area approached the same level of loss (U.S. Bureau of the Census, 1982). The satellite cities facing Manhattan on the west bank of the Hudson River in New Jersey also declined. Factories were closing, White middle-class people were moving to the suburbs further west, and local businesses found difficulties in competing with businesses located in Manhattan. During this time some 80,000 Cubans settled in Union City and adjacent places, creating the second largest Cuban community in the United States. Union City offered housing suited to the internal diversity of the Cubans. It had many old walk-up apartments with modest rents, family homes with average comfort, and even mansions that could be renovated. Light industry installations were attractive to Cuban entrepreneurs and the closeness of suburban areas facilitated an outreach of retail activity related to local businesses.

At first, the Cubans caused concern and resentment among the long-established natives, but with time, local leadership conceded their positive role in reversing the economic decline and rehabilitating the city. Typical of the new era are development plans for nearby Weehauken and West New York involving a substantial investment in renovated and newly constructed housing, transportation, office, and commercial facilities (Dorf, 1984). In Union City, Cubans predominate among new immigrant groups and have managed to develop a community showing clear signs of social and economic mobility.

In Jersey City, other nationalities are doing the same kind of thing. Where professionals of Jewish, Italian, and Irish descent once practiced, there are Filipinos, East Indians,and Taiwanese. People who speak Arabic have formed an ethnic community in the Journal Square area once inhabited by white-collar commuters of European origin. Koreans own or manage some of the shops and services in the middle-class strip a block away from Kennedy Boulevard. Certain signs of progress can be observed in the lower-class neighborhoods marginal to the strip, where native Blacks, Puerto Ricans, and some Hispanic immigrants live. However, most of these people seem to be factory and service workers or unemployed. In brief: The new immigrants have partially replaced the White European population that made up the middle and upper-middle strata of the local social order. With the stamina typical of people in a challenging situation, they seem to have set about improving Jersey City as their place of settlement. Of course, not all the new immigrants share in this comparative affluence. Many immigrants suffer from downward mobility; still others have problems in adaptation and personal limitations. Then also, immigrant groups carry with them certain class distinctions from their nation of origin, which limit mobility in ways internal to each group.

Except for scattered appearances in the four community types just described, Chicago's new immigrant elite live in such suburban places as Glenview, Lake Forest, Skokie, Wilmette, and Winnetka, Illinois (Kiefer, 1984: 130-133), where their homes are interspersed among native Whites. In the greater New York metropolitan area, affluent new immigrants seem to live interspersed among native Blacks who have suburbanized, preferring this to central-city succession neighborhoods of equivalent quality. In the various municipalities of Nassau County, Long Island, it is hard to determine whether the succession is led by native Blacks or new immigrants or by a combination of both. In any case, the new immigrant presence is unmistakable. Thus it merits consideration as a *fifth community type, or scattered suburban settlements ranging from lower- to upper-middle-class standing,* according to American majority standards.

RESEARCH METHODS

The typology just presented has practical implications for gathering data on new immigrants. By focusing on relevant geographic areas, simplicity and efficiency can be attained in the methodology needed for success in reaching our primary objectives of revising paradigms of urban composition and change, and enabling new immigrants to use the data produced for participation in public policy. In one way or another, short-range strategies for collecting demographic information must rely on a survey design aimed at obtaining a valid and reliable recording of variables unavailable from conventional sources.

The sample frame is a particularly troublesome aspect of such an endeavor. The general lack of credible demographic information on new immigrants at local levels usually motivates the average researcher to opt for a citywide or even metropolitan sampling plan based on area probability selection. But this approach tends to be complicated and inefficient, unless the new immigrant portion of the research is part of a large-scale sample of the total population. Huge expenditures are necessary for screening respondents and little or no analytic refinement can be obtained about community areas or neighborhoods. Measures of demographic changes thus cannot be effectively related to the social and economic context in which they are taking place.

A preliminary study involving observation and purposive interviews can help in focusing the sampling on urban segments identified as including the majority of new immigrants, as earlier demonstrated in the typology. Somehow a measure of external validity must be gained in developing a sample along these lines. Part of the problem is solved by assembling whatever general demographic information can be obtained and using it as a stratification guide for checking the composition of the respondents selected at random. Special attention must be given to age, gender, and household composition. Age and gender are crucial to any match with the total population, and they can be used to estimate vital rates and reconstruct the total population with greater confidence. Because of the wide variety and sometimes temporary nature of living arrangements among new immigrants, household composition must be measured as accurately as possible. Researchers must avoid missing people who are genuine residents but who may be skipped by conventional surveyors: These include undocumented immigrants and sojourners undecided about whether to return to their native countries.

The ultimate test of external validity depends on the conducting of repeated surveys in which the age, gender, and household composition of the new immigrant population can be compared at periodic intervals. A longitudinal design is also essential to assure reliability in the measurement of such variables as marital status, fertility, and migration, which are central to the study of change in any population. Another important advantage is that a series of surveys with a brief questionnaire can be adapted to produce information on how community areas and neighborhoods change in relation to the presence of new immigrants. A questionnaire of expedient size and comprehensive to all household members also helps to shift the focus of attention from a special household member who may be viewed as the source of information to outsiders, by traditional cultural norms. Finally, the approach suggested allows the use of the telephone for formal data collection instead of the traditional home interview method. This reduces the cost of data collection to a fraction of the expenditures needed for the canvassing method and seems to improve the quality of the data gathered by providing an anonymous, factually oriented environment.

Many researchers struggling with the lack of data from conventional sources and a diversity of interests in a new immigrant community are tempted to include every imaginable item in a survey. This typically results in a one-time data collection effort of gigantic proportions, producing a great deal of information of questionable value. Often the survey becomes a learning process in which the practical knowledge gained does not help improve further data collection. This problem can be solved in a longitudinal design by supplementing the standard demographic items with a modest number of questions on special topics. Periodic assessments of this kind can help maintain community involvement in the survey and enable the researchers to demonstrate the results in a more expedient manner.

In the selection of interviewers, it should be remembered that the ethnicity, gender, age, generation, accent, and social class of the interviewer can influence the possibility of access to a given household and the quality of the data collected. This applies to both the traditional canvassing method and the telephone interview situation. The validity problems that create the undercount and error in official data collection activities are related to absence of contact with coethnics and high rates of refusals and nonresponse. Therefore, researchers should give careful attention to the social aspects of the interview situation and choose the interviewers accordingly. The experience of the Latino survey in Chicago showed that carefully selected coethnics enhance the quality of data collection and reduce the refusals and nonresponses to minimal levels. The principal factor seems to be the ambience of mutual understanding and confidence established by coethnics. Whatever response is elicited by publicity before a survey and the sponsorship of new immigrant organizations may be better maintained by coethnic interviewers: They may also check, edit, and organize data for analysis—tasks that demand language skills and insight into the respondents' way of thinking.

New immigrant control and ownership of the research is strongly supported by various considerations. The most essential stems from the ineffectiveness of official data sources: New immigrants are responding to this form of inequality by creating alternative systems of data production. It would not serve this objective nor the ultimate goals of new immigrants as constituencies in the American political arena to lose or delegate power over information that defines their social and economic status and helps shape public policies to solve their problems. Beyond technical assistance, general control of the research and its dissemination and publication should remain in the control of the new immigrants.

THE RESEARCH ENVIRONMENTS

The new immigrants have arrived during times of intensified self-reliance among constituencies and technological innovations that have ushered in high-speed and extremely flexible information systems. Many native constituencies have responded by organizing and controlling data production to supplement or replace the statistics generated by official data sources. A multitude of consultant and computer firms are able to satisfy research expectations that cannot be fulfilled quickly by a census-taking process that was legally intended simply to count the population for apportionment of congressional districts and governmental program objectives. It seems reasonable to expect that foreigners who have been selected for entry to the United States by educational preference will tend to follow the example of the more progressive segments of the native population, particularly in terms of demographic data production and analysis. In so doing the new immigrants would join an environment that is rapidly becoming a pluralistic system of data production.

This form of entrepreneurship by immigrant community organizations or businesses can provide timely and accurate information immediately useful in determining public policy at federal, state, and local levels. This is particularly crucial, given the notably increased competition for reduced benefits from public programs. The relative success, therefore, of a particular nationality group will likely depend on its ability to develop and control data production systems in step with native segments of the U.S. population.

Another avenue suggests itself in the social history of the American demographic establishment. At the Census Bureau, for example, the children of "old" immigrants eventually occupied many of the managerial positions below the level reserved for appointment by presidential administrations. Their leadership evolved in an organizational environment marked by routinized adherence to established procedures and the production of standardized numerical information. Joined with the extraordinary growth and forcefulness of official statistics, this converted the Census Bureau into a foremost agent of assimilation to the American system. It seems reasonable to expect that new immigrants oriented to the same ideological position will seek succession to these positions of leadership in ways similar to their residential patterns in middle- and upper-middle-class urban areas.

The strategy of leadership succession from within is given credence by some of the long-standing practices of the Census Bureau and the American demographic profession. Both took important leadership roles in the reorganization and increased efficiency of official statistical systems in nations supplying the new immigrants. This outreach effort has been of significant magnitude. It can be argued that it elicited motives for immigration similar to those attracting other professional groups. To begin with, data specialists from Third World nations typically acquire their knowledge in the United States or in American-sponsored programs. If new immigrants with skills related to demography also acquire a sense of social responsibility to their communities, a trade-off may be possible. In exchange for the homogenizing occupational functions of assimilation, they can provide greater sensitivity and attention to the cultural patterns and needs of the new arrivals.

A different scenario emerges from recent policy redirections at the Census Bureau. Generally speaking, this appears to be a move toward the model followed by equivalent governmental organizations in many other nations. Data collection and analysis activities would be reduced to the most essential functions. If continued, this would result in specialized kinds of data production being eliminated, particularly those that require innovative and localized strategies. Along with a variety of research functions acquired over years of nurturing by academics and international developers, the task of producing data for new immigrants may be assigned to the private sector by default. This possibility is favored by limitations inherent in any nation's official data production system. Perhaps we are witnessing the decline of governmental functions that can now be more effectively performed by major corporations providing research products with greater speed and flexibility. With time and experience, these providers may enlist the talent of new immigrants and eventually acquire greater expertise than the Census Bureau on the new immigration, and also a more advanced technology for data production.

Whether new immigrants assume control of data production in the pluralistic model, seek succession in the governmental demographic establishment, or serve as employees of corporate data producers, the information gaps are likely to be filled somehow in the near future. An issue such as the revision of paradigms of urban composition and change touches on the interests of many segments of the American public in addition to the new immigrants themselves. As often happens, researchers not belonging to the population researched may take the initiative and develop this subject matter as a major topic of intellectual concern. Although this would represent a significant loss of power to new immigrants, the receiving metropolitan areas would be benefited by the availability of information of use in the formulation of more coherent and effective policies for the future.

The research strategy proposed seems most appropriate at present for any and all of the alternative research environments. By developing a simple but rigorous methodology, basic questions about the new immigrant population and its relation to urban ecology can be answered as a short-term and intermediate strategy to fill the most essential information gaps. In the process, the immigrants of today will have attained at least one attribute that justifies their designation as "new": involvement in an innovative data production system for local urban areas. This is bound to be one of the most exciting chapters in the history of immigration to the United States and in the literature on paradigms of urban composition and change.

Part II

Institutional Contexts and Responses

Race, Color, and Language in the Changing Public Schools

RICARDO R. FERNANDEZ
WILLIAM VELEZ

ETHNICITY AND EDUCATION

Perhaps no other institution is as central for Americans as school. So far the eighties have seen a renewed interest in education and a growing debate on how to reform the schools, as exemplified by the National Commission on Excellence in Education's *A Nation at Risk* (1983). To understand the ideological sources of the present situation, we have to look at the history of schooling.

From its start, the "common school" was tied in with efforts at creating an "American" identity. With these efforts there was a belief that some foreigners make better Americans than others, as reflected in earlier immigration quotas. However, this melting pot ideology co-existed with a decentralized structure of most school systems during the nineteenth century. This allowed immigrant ethnic groups to put political pressure on school systems to secure at least some recognition of their special needs and to resist biased learning materials.

In the early twentieth century, in the name of "efficiency," schools became more centralized and bureaucratized, and control came to rest almost absolutely in the hands of "experts." It is not clear, however, whether or not this was designed to obliterate ethnic diversity. Certainly there was a desire to produce enough assimilation to make the newcomers acceptable to the native population.

Gradually, more favorable attitudes toward cultural variety have emerged. This is as much a recognition of the persistence of cultural differences as it is the end result of pervasive segregation of minority students, who were never targeted for complete assimilation, especially at the socioeconomic level. Increased tolerance, however, does not mean embracing cultural pluralism.

In the mid-1980s we see the major urban school systems serving a clientele that resembles the people they faced three to four generations ago. These are the disadvantaged native minorities, who have become numerical majorities; the new immigrants, both legal and un-documented; and the children of the refugees.

This chapter attempts to give an overview of the many issues surrounding urban school systems in this new era. Our first task is to assess the fluid meaning of the ethnicity in urban America. Related to this are demographic profiles of cities and school systems in the major urban metropolitan areas. Next we will review the evidence on the achievement of ethnic/racial minorities in urban schools, with special emphasis on the causes of school leaving. We will discuss some school practices such as tracking and segregation and how they affect minority students.

The section on bilingual education will describe school programs aimed at meeting the needs of language-minority children, who are found increasingly in large metropolitan areas, such as Boston, Los Angeles, and Chicago. Ongoing trends that crystallize some of the reform efforts, such as the "effective schools" movement and the growth of minimum competency testing, will also be analyzed, along with other issues.

FLUID MEANING OF ETHNICITY

Our first task is to clarify the fluid meaning of ethnicity. The term "ethnic group" implies some degree of solidarity, and assumes that "its members are at least latently aware of common interest" (Petersen, 1980). There are various criteria of ethnicity that are relevant in the American context: race, language, national origin, and religion. The emphasis on assimilation in the United States has meant that Anglo-Saxon Protestant cultural traits are preferred. To the extent that successive waves of immigrant groups gradually lost their cultural distinctiveness, their ethnicity has become more symbolic than real. In other words, the survival of some customs and culinary preferences sustain personal identity in the private sphere of life, with no impact on the life chances of the individual.

By contrast, in our usage, the term "ethnic groups" refers to people who live at the bottom of the socioeconomic structure and who face institutionalized barriers in their efforts to achieve social mobility. These are groups that are classified as racial minorities—Asian Americans, Native Americans, Pacific Islanders, Black Americans—as well as national-origin groups that have suffered discrimination in the past and still have low prestige and socioeconomic status—Puerto Ricans, Mexican Americans, Cubans, and other Spanish origin groups we conveniently label "Hispanics." They also include a growing number of refugees from non-European nations. Due to their young average age, they are the most affected by educational institutions. Because of their special needs, they are highly dependent on the available school resources. Even though the official categorization of some of these groups has vacillated between an ethnic and a racial group (as for Mexican Americans and Puerto Ricans) the common experience of institutional discrimination sets them apart from the dominant population.

Currently, the official position of the Bureau of the Census is that Hispanics can be of any race, assuming that they are primarily national-origin groups joined together by a common Spanish ancestry. Ironically, half of all Americans who identified themselves as being of Spanish origin in the 1980 census refused to check off any of the racial categories offered by the census and instead wrote in mostly Hispanic subclassifications under the "other" category (Tienda and Ortiz, 1984). This was especially true of Puerto Ricans and Mexican Americans, the most economically disadvantaged of the Hispanic subgroups. We understand this as evidence of a transformation of the concept of ethnic groups and its relevant indicators as converging with the status of a minority group. It is not by chance, as we explain below, that this new meaning of ethnicity coincides with the "browning" of America.

One in five Americans today is Black, Hispanic, Asian, or other minority (Kasarda, 1984). Ethnic/racial minorities are highly concentrated in urban metropolitan areas: 88 percent of our Hispanics live in such areas; 57 percent in central cities. New York City's population is 52 percent minority persons, and in many large cities, such as Detroit, Chicago, Atlanta, and St. Louis, minorities are the majority (Kasarda, 1984). Given their high birth rates and disproportionate share of both legal and illegal immigration, they are becoming a numerical majority in the largest metropolitan areas. For example, in Los Angeles, minorities now constitute 52 percent of the city's population, with one in ten persons being Asian, and one in three Hispanic. Even where they are not a numerical majority of the general population, they often constitute a majority of school enrollments due to higher fertility and their youth. In

TABLE 5.1 Largest Urban School Systems: Racial and Ethnic Characteristics of Cities and of Students

City	Population[a]						Student Enrollment	
	Total	White[b]	Black[b]	Hispanic	Other[b]	Percentage Minority	Total[a]	Percentage Minority
New York	7,072	3,669	1,694	1,406	303	48	924	74
Los Angeles	2,967	1,419	496	816	236	52	544	78
Chicago	3,005	1,300	1,188	422	96	57	436	84
Dade Co. (Miami)[c]	1,626	754	270	581	21	44	222	71
Philadelphia	1,688	964	634	64	28	43	208	73
Detroit	1,203	402	754	29	18	67	203	89
Houston[d]	1,595	834	436	281	43	48	189	77
Dallas	904	514	264	111	15	43	128	74
Baltimore	787	342	428	8	9	57	120	80
San Diego[e]	876	602	76	131	67	31	111	48
Memphis	646	332	305	5	4	49	105	77
Washington	638	164	445	18	11	74	92	97
Milwaukee	636	454	146	26	11	29	86	58
New Orleans	558	225	305	19	9	60	81	90
Columbus	565	428	124	5	8	24	69	36

City							
Albuquerque	332	199	8	112	12	40	73	47
Atlanta	425	135	280	6	3	68	68	93
Fort Worth[d]	385	245	87	49	5	37	65	57
Denver	492	327	58	92	15	34	63	61
San Francisco	679	355	85	83	156	48	60	83
El Paso[d]	454	151	17	279	6	67	60	74
San Antonio[d]	786	298	57	422	9	62	60	90
Long Beach	361	245	40	51	26	32	56	53
St Louis	453	239	205	6	3	47	56	80
Austin[d]	380	264	43	67	6	31	56	48
Boston	563	382	122	36	23	32	56	70

SOURCE: Council of the Great City Schools (1983), California Department of Education (1982), California Department of Education, Bilingual Education Office (1984a, 1984b).

a. In thousands.

b. Non-Hispanic.

c. Dade County includes the cities of Carol City, North Miami Beach, North Miami, Westchester, Hialeah, West Little River, Miami Beach, Miami, Coral Gables, Olympia Heights, Kendale Lakes, and Kendall.

d. Enrollment figures are for the 1983-1984 school year.

e. Enrollment figures are for the 1980-1981 school year.

Milwaukee, for example, where minorities constitute only 29 percent of the city's population, about three out of every five students in the school system is a member of a racial or ethnic minority (see Table 5.1). In Chicago, the 1983 Student Racial/Ethnic Survey revealed the following breakdown in student enrollment: Black, 60.6 percent; White, 15.6 percent; Hispanic, 21.2 percent; Asian or Pacific Islander, 2.4 percent; Native American, 0.2 percent. Similarly, 1982 data for Dade County (Miami), show the following: Black, 31.3 percent; White, 28.8 percent; Hispanic, 38.8 percent; Asian, 1 percent. Currently, 22 of our 26 largest city school districts are "minority-majority" (see Table 5.1).

The future of the major urban school systems is clouded by the relative economic decline of most of the largest cities in the United States. A disproportionate share of investment capital and jobs is flowing to the suburban belts around cities as well as to rural areas. "White flight" from desegregation, coupled with suburban growth and movement from the Frostbelt to the Sunbelt, has exacerbated the problem. With a shrinking tax base from which to fund public services, many school districts are already on the verge of fiscal insolvency. As this chapter was being written, Chicago teachers were on strike. Their unwillingness to accept cuts in their fringe benefits brought them into conflict with the school board, which claimed these cuts were necessary to keep the system afloat.

There is mounting evidence that many of these school systems have failed to provide a quality education for all students and, in particular, for minority students. In the same manner in which they failed to adjust to European migrants (Greer, 1972), urban schools are failing to keep and to graduate minority students at acceptable levels, as we will see when we look at their record.

URBAN EDUCATION FOR MINORITY STUDENTS

In a society where educational credentials are a ticket to better paying jobs, fewer and fewer minorities are completing their schooling. The dropout rate from high school for 18 to 19 year-old Hispanics was 36 percent in 1981; this was more than double the national rate. In California only about two-fifths of Mexican American males, 25 years of age and over, completed high school, while approximately two-thirds of comparable Whites finished high school (Ford Foundation, 1984). Orfield et al. (1984) report that in Chicago, in 1980, 31.3 percent of

Whites, 49.4 percent of Blacks, and 56.7 percent of Hispanics 25 years or older had dropped out of school.

A major reason behind these high dropout rates for Hispanics is that they experience high rates of grade-level advancement problems. A recent study of Hispanic teenagers found that one out of four students of Mexican and Puerto Rican origin were two years or more behind their classmates (Olivas et al., 1980).

The question arises whether this failure can be attributed to the schools or is a result of a lack of preparedness due to the poverty and special problems minority children bring to their schools. It is a well-known fact that Hispanics, Blacks, and Native Americans are at the bottom of the income structure in the United States. In 1982, for example, 30 percent of the Hispanic population was classified as having incomes below the poverty level—twice the figure for the total population. Add to this the language adjustment faced by the new-comers (especially Hispanics and refugee children), and the situation gets complicated. A recent study of a national sample of high school students of Spanish origin found that it could not explain much of the variance in school delay (National Center for Educational Statistics [NCES], 1981). Since the predictors included socioeconomic status, English and Spanish proficiency, and length of residence, the findings suggest we can rule out input factors as the major cause of school failure among minority children.

The authors suggest that "school- and system-level discretionary policies and practices are the major causes of school delay" (NCES, 1981: 75). Thus, for example, Mexican-American students have higher delay rates in Texas than in California because school authorities there are more likely to force a student to repeat a grade (Carter and Segura, 1979). This is just one of many school practices with deleterious effects for minority students. We expand on this subject in the following section.

SCHOOL PRACTICES AND STUDENTS OF COLOR

Even though many schools have become desegregated, there is a mechanism for resegregating within the same school: tracking. The curriculum a high school student selects has important implications for future academic options. Academic (college-oriented), general, and vocational curricula are designed to prepare students for very different futures. Because only those students in the academic track take college-oriented courses, a disproportionate number of minority students will never attend college. While two out of every five White high school seniors are in a college preparatory curriculum, one out of three Black students and only one out of every four Hispanic students are enrolled in the same track (NCES, 1980). Students in noncollege curricula take fewer courses in advanced math, a good predictor of college entrance and completion (Velez, 1983). But even when noncollege-track students make it to college, and even after controlling for number of high school math courses taken, college track students outperform them and graduate at higher rates (Velez, 1984).

In many urban school districts some schools have been designated as specialty or magnet schools, skimming off the brightest from the rest of the system in an effort to create a systemwide academic track. At the same time, other schools stop offering even the minimal coursework necessary for college. For example, Orfield and associates (1984) found seven high schools in Chicago that made it quite improbable that their students would enter college because they lacked adequate curricular offerings in science, math, and foreign languages. All these schools had predominantly minority enrollments. It is not surprising, then, that Hispanic and Native American students have lower college aspirations and lower self-assessed ability to complete college than do white students (NCES, 1980).

Another school practice affecting Hispanic students is segregation. The period of 1968 to 1980 saw a decline in the segregation of Black students in every region but the Northeast. During the same time, however, the segregation of Hispanic students increased nationwide (Orfield, 1982). This increase was particularly high in the West and in the states like New York, Texas, and New Jersey, where most Hispanics reside. This is of great relevance since the quality of instruction is superior in schools having White students (Carter and Segura, 1979). But even if they attend integrated schools, Hispanic students may be segregated by classroom assignment patterns. Thus English as a second language courses and bilingual programs are used in ways that lead to segregation within the school. Ironically, some school districts have at times conveniently labeled Hispanic students as either white or minority in their efforts to get around court-ordered school integration. Reassigned to many different schools, the ones who really need bilingual instruction are now too few to justify programs that meet their unique needs. A more detailed analysis of how school desegregation has affected language minority students will be given later.

Another school practice that seems to affect Black students negatively is that of disciplinary actions, such as suspensions and expulsions. In the fall of 1978 Black students in the nation's public elementary and secondary schools were disciplined at twice the rate their enrollment proportions would have indicated (NCES, 1984). This suggests a serious problem of maladjustment that is being met with punishment instead of specialized help.

Yet another practice in some of the largest urban school districts is overcrowded classrooms. Recently, the chancellor of New York City schools admitted that about a third of all high school classes still exceed the maximum 34 students required by the School Board (*Education Week,* December 5, 1984). It is well established that reduced class size is associated with increased academic achievement (Glass and Smith, 1978). In Chicago, mobile units are used in schools with high minority enrollments to alleviate overcrowding. The solution, however, resulted in increased segregation of minority students. In 1981, there remained 320 mobile units in Chicago (from a high of 1300 in 1962). More than half of these (167) were located in 28 segregated (Hispanic) schools, with 98 mobile units in 17 segregated Black schools (Cruz, 1981).

This necessarily brief review highlights many of the practices and conditions that exist in urban school districts for minority students, even when they constitute a majority of the student enrollment. The following section offers a more detailed account of several key issues and major trends that affect minority students generally, and language minority students in particular. They were selected because of their significance and impact on educational outcomes.

MINORITIES IN SCHOOLS: KEY ISSUES AND MAJOR TRENDS

BILINGUAL EDUCATION

There were approximately 2.4 million children of limited English proficiency (LEP) in the United States in 1982 (U.S. Department of Education, 1984). Although more than a dozen states require programs for LEP children (in spite of federal funds for bilingual education, which in 1983 served 234,000 students), it is estimated that between 1.5 and 1.8 million children remain to be served (Orum, 1984). Beginning in the early to mid-1970s, the influx of thousands of children—of undocumented workers, mostly from Mexico; millions of immigrants; and more than 705,000 refugees from Southeast Asia, Central America, the

Caribbean, Mexico, and other parts of the world—has altered the character of the school population in the United States. Special language-assistance programs can be found in every major urban public school district across the country. They provide extensive bilingual education and English as a second language programs for pupils who come from backgrounds of Spanish, Vietnamese, and several other Asian and European languages. Based on data from the U.S. Department of Education's Office for Civil Rights for 1980-1981, significant LEP enrollments were found in Boston (with 12,867 children—19 percent of the total), Los Angeles (with 101,795 children—19 percent of the total), New York (with an estimated 60,000 children—6 percent of the total and San Francisco (with 12,235 children—21 percent of the total). By spring of 1984, the figures for Los Angeles had increased to 127,185 children (23 percent of the total) and 17,516 children (29 percent of the total), respectively. This means that new arrivals continue to settle in these large urban areas, contributing to the "browning" of urban America.

Since the mid-1960s, Hispanic groups have led the struggle to make instruction understandable for language minority children. Indeed the one item that has always ranked high on the educational agenda of all Hispanic groups, regardless of social class or political orientation, is bilingual education. Recent national surveys (Yankelovich, 1984; Coca Cola, 1983) confirm the importance of Spanish for all Hispanics. They appear desirous of retaining their language and culture more than previous immigrant groups. This attitude shows clearly in the educational arena through the almost unanimous support for bilingual education.

The fight for bilingual education has affected every major urban school system, and can be attributed to the growing political influence of Hispanic as well as other language minority groups. However, it has focused attention on only a fraction of the nation's public school students—those with limited skills in English. In doing so, the equally urgent needs of the rest of the language minority students who do not qualify for special programs have generally been ignored. In the case of Hispanics, these are students who are English dominant or at least proficient enough to be part of the all-English program. Although they constitute a majority of the Hispanic student population in large urban districts, few services and programs are targeted for them. For many administrators, to serve Hispanic students means only to provide bilingual education. Fortunately, as other issues besides bilingual education are discussed, the *total* educational needs of Hispanics are more likely to be addressed.

Effective Schools. Because of the recent debate on excellence in education, effective schools, and the not-so-recent debate on the outcomes of desegregation on students, the focus of attention has shifted to a broad concern for academic achievement by *all* students. It should also be noted that, in spite of the large increases in minority enrollment in the public schools, the debates have been conducted without any real input from language minority populations. Data continue to be reported by school districts and state education agencies in broad categories, such as "minority." When minorities are dis-aggregated into subgroups—for example, Asian, Hispanic, Native American, Black—typically there is no breakdown provided by national origin—Chinese, Filipino, Vietnamese, Hmong, Laotian (Asian), or Mexican American, Puerto Rican, Cuban, Dominican (Hispanic). Significant differences exist among these groups, and aggregate figures tend to conceal them.

Achievement data generally reveal that Hispanic students tend to perform below the national level in reading, although some gains have been made from 1974 to 1980 (National Assessment of Educational Progress, 1982a). Similar data for students from homes in which English is not the dominant language indicates that these youngsters "generally performed below students for whom English is the dominant language on the 1979-1980 national reading/assessment." Another important finding was that

> language dominance has different effects for people in different schools and socioeconomic strata. In general, OL students [i.e., students from homes in which a language other than English is often spoken] attending advantaged urban and private schools, those coming from homes with many reading resources, and . . . those having a parent with post-high school education performed near, at or above national levels. OL students attending disadvantaged urban schools, those coming from homes with few reading resources, and those whose parents have not completed high school performed considerably below their more advantaged peers [National Assessment of Educational Progress, 1982b].

This last finding explains the educational success of the first Cuban migration to Miami in the early 1960s, and the high achievement rates of some immigrant children (Chinese, Japanese, Korean, Filipino) and some refugees (mostly Vietnamese), in contrast to the abysmally low achievement rates of others (Hmong, Laotian, Samoans, Chamorros [Guamanians], Haitians, Mexicans, Salvadorans, Guatemalans, and the second wave of Cubans who came in 1980). In Wisconsin, where 44 percent of male Hmong heads of household report less than a fourth-grade education, the 1980 census reported a much higher educational attainment rate for Asians based on earlier arrivals (Vietnamese, Chinese, Japanese, Indian) who were predominantly professionals. Achievement rates for children from these families reflect much higher scores (even though the children were LEP when first enrolled in school) than for pupils from low socioeconomic levels. The implications of these data are significant: Aggregate data for certain language minority groups, for example, Asians or Hispanics, can be deceptive. They must be disaggregated by subgroup in order to reduce the distortion and present their full meaning.

Another indicator of low educational attainment for Hispanics can be seen in the number of high school graduates, which increased from 832,000 in 1975 to just over 1 million in 1980. However, this increase in absolute numbers actually represents a decline in the proportion of Hispanics who had completed high school—from 57.5 percent in 1975 to 53.7 percent in 1980 (NCES, 1983).

In spite of their numbers, language minority children have been ignored by researchers in the effective schools movement, which has focused almost exclusively on the achievement of Black children. The danger here is that remedies may be framed with one minority group in mind (Blacks), and prescribed for *all* of the children without taking into account the unique curricular needs of language minority students.

MINIMUM COMPETENCY TESTING

The response of most state legislatures to the report by the National Commission on Excellence in Education (*A Nation at Risk*; 1983) has been to enact laws that establish minimum competencies and standardized testing procedures to determine whether these levels are being met. Of particular concern here is the use of achievement tests to determine grade promotion and, ultimately, the graduation of minority students who in the past have been labeled as underachievers because they tend to score lower than Whites in these tests. To be sure, the movement toward minimum competency testing (MCT), with its rush to establish standards, raises the prospect that minority children, particularly language minority children, may suffer disproportionately the negative impact of these new educational requirements. In order to grasp the speed of the response to MCT proposals across the country, it should be noted that by mid-1984, 48 states were considering new high school graduation requirements, and 35 had already approved such changes as more time in the school day, or more days in the school year, or more time of actual instruction on a daily basis (U.S. Department of Education, 1984b). Other questionable school practices, such as ability tracking based on test scores, may begin again in the name of the pursuit of excellence, much to the detriment of minority students. This possibility was made more real by the April 1984 decision of the Court of Appeals for the 11th Circuit in *Debra P. v. Turlington* that the State of Florida's MCT program could continue, even if it affects Black students disproportionately. As of March 1984, some 1100 students in Florida had been denied diplomas based on the test, and over 620 of them were Black (Schwartz, 1984).

The institution of MCT has been done without regard to the consequences for racial and ethnic minorities (Siegel, 1983). An analysis of the *Texas Assessment of Basic Skills* conducted by Intercultural Development Research Association (IDRA) in 1981 revealed a "significant lag in mastery of basic skills by some groups of students (i.e., ethnic minorities, low-income students, migrant pupils, and students with special learning needs) and this lag increases as the students progress through the state's school systems" (Cortez, 1981: 1). The potential impact of widespread MCT on language minority children is ominous indeed, especially if one considers that it will probably lead to the withholding of regular high school diplomas for thousands of youngsters. Instead, they will receive "attendance diplomas" that will foreclose for them any real possibility of a postsecondary education. Veiled under the aura of objectivity, MCT legitimizes the exclusion of groups of children from effective participation in higher education. Eventually this forecloses any chance of aspiring to the better jobs. How LEP students will fare is even more doubtful, and it is certain that legal challenges will be made on behalf of these children to prevent the denial of regular diplomas based on their scores on English language tests. Given these realities, Orfield et al.'s (1984: 123) assessment of conditions in Chicago may be a premonition of what lies ahead for minorities in major urban school systems:

> What we have in Chicago, then, is a two-tiered system of public secondary education consisting of dead-end to middle-of-the-road schools, and the advanced training schools. The schools on the bottom offer students almost no chance to seriously prepare for competitive higher education.

SEGREGATION/DESEGREGATION

Historically, the problem of racial segregation has been viewed as a Black/White issue in American society. This is understandable given the experience of slavery and its aftermath in the twentieth century. Nevertheless, segregation of Asians in California and of Chicanos in Texas and California in the 1930s, 1940s, and 1950s was quite common, and there were a number of successful legal challenges to segregatory practices by Mexican American plaintiffs (Fernandez and Guskin, 1981). By 1970, federal courts had determined that, for the purpose of school desegregation, Mexican Americans should be treated as an "identifiable ethnic minority group." In 1973 the U.S. Supreme Court found in *Keyes v. Denver* that in the Southwest Chicanos had been discriminated against and segregated in a manner similar to Blacks. These decisions had an impact in Texas and in other states where school districts had been classifying Chicanos as White for the purpose of mixing them with Blacks and calling those all-minority schools "desegregated." After the Denver case, the concept of triethnic desegregation began to be used by courts and planners in framing remedies. However, a problem that arose immediately was that desegregation plans were being drawn solely on the basis of racial and ethnic criteria without regard to the linguistic needs of language minority pupils. Assignments in those early days usually did not take into consideration that LEP students needed to be clustered in schools and classrooms in sufficient numbers to allow viable bilingual and other language-assistance programs. An important precedent was established in the Boston desegregation case, where Hispanic (mostly Puerto Rican) and Chinese students were able to receive home language instruction in a desegregated setting under a plan approved by the federal court, which assigned LEP based on their language needs. The underlying principle had been stated earlier in a Texas case: "No remedy for the dual system can be acceptable if it operates to deprive members of a third ethnic group of the benefits of equal educational opportunity" (*U.S. v. Texas Education Agency (Austin),* 467 F. 2d at 848).

The problem of segregation for language minority students, especially Hispanics, appears to be increasing. An analysis of trends published by the Joint Center for Political Studies in Washington, D.C., based on longitudinal data from 1968 to 1980, pointed out that while the segregation of Black students declined significantly during the 1970s, especially in the South, Hispanic segregation increased during the same period, particularly in areas with large and growing populations, such as the West and Southwest (Orfield, 1982; Orfield et al., 1983). Even the most cursory review of the changing demographics of urban school districts points to increasing concentrations of minority students—

Blacks and language minority—across the country. Usdan (1984) points out, for example, that about two-thirds of the Hispanics in the United States are found in only three states (California, New York, Texas), while approximately 85 percent of all Hispanics live in only eight states (Colorado, Florida, Illinois, Michigan, New Jersey), plus the three mentioned earlier. Given the current birth rates of Hispanics, and their relatively low median age (over 30 percent of the Hispanic population is between the school-going ages of 5 to 17, as opposed to 23 percent for the general population (Usdan, 1984), it is likely that the problem of segregation, that is, racial/ethnic isolation, will worsen for Hispanics and other language minority students. In fact, one could say that the segregation of Hispanics has yet to be dealt with in American schools and cities. However, of necessity, this will change due to the projected increases based on current population growth rates.

DROPOUTS

It is well documented that minority students are more likely to leave school before graduating from high school. Most studies that focus on language minority students have addressed the needs of Mexican Americans and Puerto Ricans (Lucas, 1971; Aspira, 1976; Aspira of New York, 1983; Santiago, 1984; Orum, 1984; Kyle, 1984; Hernandez, 1983), but researchers are also investigating the conditions of other language minority students (Steinberg et al., 1984). The definition of what constitutes a dropout accounts for the large disparity in the rates reported for various groups, which range from a low of 10 percent to a high of 25 percent for Whites and a low of 25 percent to a high of 70 percent to 80 percent for minority students. For example, Steinberg and associates (1984) report a dropout rate for the general population of 25 percent, which is the same for Black students. For Hispanic students, however, they note a significant increase from 30 percent in 1974 to 40 percent in 1979. In 1971 Lucas had reported a 71.2 percent dropout rate for Puerto Ricans in the Chicago public schools, a finding that Kyle (1984) has corroborated. Aspira of New York's report indicates that eight out of ten Hispanic students in the city's public schools who enter the ninth grade fail to complete the twelfth grade, as compared to seven out of ten Blacks, and five out of ten Whites (1983).

The causes of dropping out are complex, yet most researchers attribute them to several factors: lack of academic success, low-income family background, low educational attainment of parents, insufficiency of the curriculum to address special language needs, and delayed education rates. One researcher (Kyle, 1984) has added a new element—fear of gangs in the neighborhood of some Chicago schools—as a direct cause of Hispanics dropping out of school. Citing 1980 data from the National Center for Education Statistics, Santiago (1984) lists worries over money, family obligations, and a lack of parental interest as factors cited by students as interfering with their school work. Orum (1984) points out that 25 percent of Hispanic children (ages 14 to 20) were below grade level, and 12.1 percent of all Hispanic (mostly Mexican Americans and Puerto Rican) sophomores in 1980 were two years below grade, as compared to only 4.1 percent of Whites. She also notes that the level of expectation for postsecondary educational opportunities among Hispanic students is much lower than for other minorities (Asians and Blacks).

What conclusions can be drawn from these studies? First, that the dropout problem has existed for a long time and is likely to continue to plague major urban school systems in the years to come. Second, and perhaps more important, that most of the studies about dropouts have a very short longitudinal range, usually four years, when it is clear that the problem needs to be studied across generations. Is it possible for a socializing agency such as the school to do for newcomers and for racial and language minorities in general what it failed to do for earlier generations of immigrants? In spite of the rhetoric of equal opportunity for all, the high rate of failure for minority students seems to confirm the hypothesis that "the failure of many children has been, and still is, a learning experience precisely appropriate to the place assigned them and their families in the social order" (Greer, 1972: 152). When viewed as a reflection of the social, economic, and political system of our society, schools appear to be very efficient, as they have always been, in sorting out students. As Greer (1972: 152) puts it:

> The fact of the matter is that American public schools in general, and urban public schools in particular, are a highly successful enterprise. Basic to that success is the high degree of academic failure among students. . . . The schools do the job today that they have always done. They select out individuals for opportunities according to a hierarchical schema which runs closely parallel to existing social class patterns. The problem today is that there is an increasing shortage of even low level employment options for those at the low end of the public school totem pole. As a result, the schools now produce people who are a burden upon, rather than the mainstay of, the socioeconomic order.

This absence of a historical perspective flaws or severely limits attempts to address the problem of the school dropout. Solutions are proposed that focus on the more visible, immediate manifestations of the problem. However, the totality of the problem is lost in this narrow delineation, which limits the parameters of the discussion on the real nature of the question. Are schools the highly idealized vehicles of social opportunity for the masses of poor children, or are they, to paraphrase Joel Spring, "sorting machines" to perpetuate the social and economic order that exists? The answers to these questions are at the heart of the matter.

FISCAL AND POLITICAL TRENDS

In analyzing the conditions of major school districts across the country, Cibulka (1982) notes several trends that create political problems for administrators and school boards. The most significant are the following: (1) declining property values in most large cities, especially in the Frostbelt, and stiff competition for tax dollars between schools and other government agencies; (2) reduced federal revenues in spite of the increased costs brought on by mandates to desegregate, and to provide services for the handicapped and students with limited proficiency in English; (3) diminished state revenues with higher expenditures in urban schools; and (4) declining public support for the public schools from an aging White population with fewer school-age children or with children in private or parochial schools, in districts that have increasing numbers of minorities.

The need for reform was discussed extensively in the 1970s, and several states (California, Texas, Florida, Colorado, New Mexico, Michigan) undertook efforts to change their systems of financing public education (Cortez, 1984). Most of these attempts at reform were in response to legal challenges mounted by minority plaintiffs. Thus, the major case in California (*Serrano v. Priest*) was brought by a Mexican American, while the most important case in New Jersey (*Robinson v. Cahill*) was initiated by Black plaintiffs. The landmark case at the Supreme Court is *Rodriguez v. San Antonio,* in which the court found that education is not a fundamental right under the U.S. Constitution, leaving it up to the state legislature to remedy glaring inequities in the Texas school financing system. The current Texas school finance case, *Edgewood ISD v. Bynum,* involves one of the poorest districts with an almost completely Chicano enrollment.

Reforms did take place in many states, mostly in the form of increases in state dollars for public education, but minimal or no decline at all in inequality was evident in any of the states mentioned earlier. Brischetto and associates (1979: 81) points out that equity considerations may, in some cases, dictate an *un*equal distribution of educational resources according to different costs and needs. He suggests that "if equity is defined as distributing resources according to need, the result might be an eventual *increase* in overall inequality." His study also found that, typically, inequality is wealth related, not need related. Further, state legislatures and courts tend to focus on ways to reduce wealth-related disparities through a variety of approaches, which include power-equalization formulas as well as increased revenues from state and federal governments.

The limitations of this chapter do not permit an in-depth examination of school finance and its impact on minority student populations in major urban districts. Nevertheless, for a simple illustration of the complexity of the problem, let us look at current events in Texas, where a new major legal challenge has been mounted against the state system of school financing. In the last several years, the Texas legislature has increased significantly the funding to public school districts, and the average per pupil aid has gone up accordingly. But these averages do not give a true picture for Hispanic students. If wealthy districts with large numbers of these students, such as Houston, Dallas, and Odessa (with a combined total of about 100,000 Chicano students) are taken out of the calculations, the revised figures indicate that almost 70 percent of all Chicano students in Texas are found in the poorest 50 percent of the districts. In fact, the gap between wealthy districts and poor districts has not been closed; it has remained relatively the same as it was before the infusion of funds (Kauffman, 1984).

Although interdistrict disparities have taken up most of the debate over school finance, an important aspect of this problem deals with *intradistrict* disparities, that is, with differences in the internal allocation of resources within districts. Weinberg (1983) noted this as a major source of inequality for minority students, who often find themselves in the oldest, poorest schools, usually located in the decaying central cities.

Espinosa (1982) used data from the Los Angeles Unified School District (LAUSD) to compare school finance and facilities between Hispanic and non-Hispanic schools during the 1980-1981 fiscal year, based on a 20 percent random sample of schools selected from the LAUSD elementary and junior high schools. The findings support two major conclusions: (1) Schools attended by children from low-income households receive disproportionately fewer funds on a per pupil basis than schools attended by children from wealthier households; and (2) schools attended by Hispanic children receive about 38 percent less on a per pupil basis and are more heavily utilized than non-Hispanic White schools in the district. Major findings in the area of expenditures were that LAUSD spends approximately $500 less per pupil on Hispanic elementary schools and $559 less per pupil on Hispanic junior high schools. When categorical funds are subtracted from the total school expenditures (since by law and regulation they should supplement, not supplant, local and state funds), the disparity between Hispanic and White schools amounts to $816 per pupil or 38 percent less for elementary schools and $732 less for Hispanic junior high schools. The White stratified schools in the study averaged 294 pupils per school while the Hispanic schools averaged 1076 students per school (363 percent larger). Generally, Hispanic schools have smaller sites and greater numbers of students, with much less playground space and physical plants that are inferior in quality; for example, they have fewer air conditioning units, less library space, and smaller cafeterias and auditoriums.

It is clear that this issue warrants a great deal of attention because of the implications for the education of minority students in the next fifteen years. In all likelihood, it will be left to minority researchers to document existing inequalities in the financing of public schools. Their work will provide the basic data on which legal challenges will be mounted and political initiatives will be launched in the coming years.

CONCLUSION

From a minority perspective, for the U.S. Secretary of Education to proclaim in 1983 that America as a nation is "at risk" because of the condition of its schools is ironic. For decades minority parents have known that the education given their children by public schools was inferior in quality to that of White children.

Even after the most obvious badge of slavery—segregation by law— was officially declared by the Supreme Court to be offensive to this nation's basic ideals, the conditions that prevailed in urban and rural schools serving Black, Hispanic, Asian, and Native American children have persisted. It took years of concerted action by Congress, the executive branch, and the courts to reverse hardened discriminatory practices against children of all colors. Judging by the evidence presented in this chapter, the battle is only partially won. Much remains to be accomplished before the goals of equity and excellence are met.

The challenge of diversity confronts American society and its institutions. The nation's economic system faces an uncertain future in a world economy that is increasingly interdependent. Dramatic social changes have taken place in the last two decades. One need but look at the composition of immigration to see that our society has changed drastically and that there will be more of a "browning" (or a "coloring") of America in the last part of the twentieth century. Garreau (1981) talks about the "nine nations of North America" in his insightful commentary on life and social changes that are visible in American society. Ernest L. Boyer, president of the Carnegie Foundation for the Advancement of Teaching, warns Americans that "we are not thinking carefully about how the growing diversity of students will affect our schools—and life beyond the classroom." The challenge is to find better ways to serve minority students because, if we do not, "the social and economic fabric of the nation will be greatly weakened." Indeed, for Boyer (1984: 5), the larger issue is "whether this nation can embrace a new generation of Americans and build a renewed sense of national unity while rejoicing in diversity. Our response to this urgent and persistent challenge will have an impact far beyond the classroom and will reach into the future as far as anyone can say."

In this state of affairs, urban public schools will have to make curricular changes. They must also recruit teachers with specialized training to deal with language minority children. Given the strong link between educational credentials and effective participation in the labor market, racial and ethnic minority children cannot afford the present rates of failure. The jobs requiring little formal education are declining in relative numbers. They are available mostly to those who have access to information and referral networks usually closed to Hispanics, Blacks, and Native Americans. Their reliance on educational credentials to obtain access to the better jobs is great and will become even greater.

Positive changes for minority students have come about in America's schools, albeit grudgingly. Desegregation demolished many barriers against Black children and teachers. Bilingual and ESL programs are now commonplace in urban districts. Physically impaired and mentally handicapped children are receiving services that previously were denied them or that were nonexistent. Girls are now enjoying many academic and athletic programs once aimed primarily at boys.

While old barriers may be gone or lowered, new barriers have sprung up that threaten the educational welfare of minority youngsters. Some of the major ones have been described above. And, heartening as the current concern over improving the schools may be after years of benign neglect or downright hostility from public officials, we find ourselves at a critical juncture in American public education. The survival of our public schools is at stake, and with it the future of present and coming generations of America's students.

6

Urban Labor Markets and Ethnicity: Segments and Shelters Reexamined

MARCIA FREEDMAN

☐ THE RELATIONSHIP of labor market segmentation and ethnicity is far more complex now than it was even a generation ago. The descendants of earlier waves of White ethnics have become occupationally integrated, while people of color, who have made progress in breaking down barriers, nevertheless remain seriously disadvantaged. In the meantime, new national and ethnic groups have become larger and more visible as a result of the 1965 changes in the laws governing immigration.

The data are not available to tease out all the interesting relationships between ethnicity and country of birth, on the one hand, and occupational mobility patterns, on the other. But there are regularities that suggest the comparative advantages available to groups with certain characteristics. In this chapter, we shall explore some critical differences in how native born Americans and new immigrants find their respective ways in the labor market and how direct competition is minimized through the development of different market structures.

Studies of actual job search reveal that a large proportion of job vacancies in the United States are filled through the operation of informal social networks (Rosenfeld, 1975; Young, 1979; Corcoran et al., 1980). This phenomenon has special significance for recent immigrants who have strong ties with members of their extended family and kinship groups. A ruling principle of the 1965 Immigration Act was family reunification, and while the act established a complex preference system based on specific relatives, newcomers have managed to bring in family members whether or not they fit official definitions of preference (Garrison and Weiss, 1977).

When it comes to employment, the same connections that make it possible for particular individuals to enter the United States also serve to get them jobs soon after they arrive. The fact that immigrants readily find work in a region suffering from chronic unemployment and underemployment, as well as in an area of economic expansion, results in large measure from their use of these personal networks. Help available from family and kin gives newcomers a competitive advantage in seeking and obtaining certain kinds of employment, particularly when the labor-market protections that serve to promote equal opportunity for native-born minorities are under attack.

From the end of World War II until approximately 1970, Americans were able to build shelters in the labor market—secure jobs with decent wages and fringe benefits adequate to counter the misfortunes of sickness and old age. These arrangements were disproportionately enjoyed by White males in the prime working ages—25 through 44— who not only held a majority of credentialed jobs, but also a majority of jobs that had shelters emanating from union contracts and the formal personnel rules of large corporations and public agencies (Freedman, 1976).

Young workers, women, and Black men were more likely to work in the less structured sectors of the market, where employment was impermanent, lower paid, and, in general, less protected. In the last decade, these sectors have grown disproportionately as the number of sheltered jobs has declined. The result is convergence, with all employment tending toward lower wage levels and weaker attachment. The steep rise in the labor force participation of women coincided with employers' realization that flexible staffing has much to recommend it in lower labor costs and the capacity to avoid personnel stockpiling. Just as contracting-out or paying overtime is preferable to increasing permanent staff, so too is the opportunity to utilize a part-time or intermittent labor force. In situations that require stability—or at least, where stability is desirable—a quasi-attached work force that values paternalism offers some employers the best of all possible worlds. That is one way that recent immigrants—somewhat in the fashion of mature women—fit employers' perceived needs the best (Bailey, 1984).

The argument outlined above requires support on both sides—with respect to labor market conditions, as well as to the employment processes of immigrants. In addressing these issues, we should keep in mind that immigrants constitute only a marginal addition to the national labor force. At the end of 1983, total U.S. population was more than 235 million. In recent years, to be sure, about one-third of annual net population increase has come from immigration, but these additions have increased the total by less than 1 percent per year (U.S. Bureau of the Census, 1984).

In any case, newcomers from abroad are seeking employment in a weak market: Major changes in the demand for labor have resulted in a reduction in the number of jobs with protective shelters, and changes in the relationships of employers and workers have resulted in the weakening of those shelters that remain. The evidence comes from permanent shifts in the distribution of employment among industries; from profound alterations in collective bargaining agreements; and from the effects of deregulation.

INDUSTRY EMPLOYMENT

The effects of shifts in employment among industries are confusing unless certain distinctions are kept in mind. One misunderstanding arises when the disappearance of jobs in the middle of the occupational structure is confused with the disappearance of the middle class. The current debate about class (see, e.g., Samuelson, 1983) is beyond the scope of this discussion, but insofar as it is stated in terms of income distribution, the place of a family reflects the combined earnings of all family members. In contrast, what interests us here is a shift in the structure of the economy as it is manifested in the long-term trend toward growth of employment in services and decline in employment in goods production.

In the 1980s, two severe recessions depressed manufacturing and construction employment, but data for the end of 1983, when the recovery was well under way, show that the downward trend was unabated. Employment in manufacturing fell from 23.5 percent of total nonagricultural wage and salary work at the end of 1978 to 20.8 percent at the end of 1983, a truly astounding decline. Put another way, while total employment grew 4.7 percent over the five years, durable goods manufacturing fell 9.1 percent, and nondurables, 4.5 percent. As construction employment fell 6.3 percent, the *share* of construction also declined, from 4.9 to 4.4 percent of the total (Klein, 1984).

These job losses were the result of well-known secular trends related to the internationalization of the economy, especially the effects of successive countries rapidly becoming able to produce more cheaply, not only textiles and apparel, but also steel and heavy construction projects. In addition, it is apparent that after decades of anticipation, we

TABLE 6.1   Wage and Salary Employment in Selected Services,
1978 and 1983

Service Sector	December 1978 (in thousands)	December 1983 (in thousands)	Percentage Change
Business and legal	3103	4466	43.9
Health	4855	5987	23.3
Education	1087	1291	18.8
Social services	973	1295	33.1
All services	16245	20041	23.4

SOURCE: Employment and Earnings (1978, 1983).

are now finally witnessing significant effects from advances in technology. These factors have adversely affected all manufacturing employment, except electric and electronics equipment and instruments among durable goods industries and printing and publishing among nondurable goods industries. The largest permanent losers are primary metals (steel) and transportation equipment (automobiles). Some large manufacturing companies posted record profits in the first quarter of 1984, but employment did not return to earlier peaks.

Among nonmanufacturing industries, wholesale and retail trade have remained steady. In retail trade, however, a large increase in employment at eating and drinking places (12.3 percent) offset losses in other kinds of business. Employment in public administration was also virtually unchanged; only an increase among state governments kept the absolute total from falling. On the growth side, finance, insurance, and real estate (FIRE) increased 15.3 percent, almost four times the overall rate. The big winners, however, were among services (see Table 6.1).

The spectacular growth of business and legal services, together with FIRE and, to a lesser extent, communications, has come about because of profound changes in production and delivery. Consumers tend to devote the same income shares to goods and services. Since the proportion of GNP devoted to services has increased, it is clear that services growth is related more to intermediate output than to final output, which is to say that the effects are more on how we run the economy than on what we actually produce (Stanback et al., 1981).

The increase in social services employment is largely in nursing homes, the result of an aging population with greater longevity. To a lesser extent, the increase also reflects more day care and other domiciliary services. The growth of the health industry, meanwhile, is an extension of longer-term trends—the introduction of Medicare and Medicaid; technological innovations that require more semiprofessional specialists; and third-party payment methods that dramatically increase administrative staffing.

OCCUPATIONAL EMPLOYMENT

The industrial pattern of growth and decline has shaped the occupational requirements and rewards of the current period. Professional and technical jobs in the services showed large gains, but so did poorly paid and unstable occupations at the bottom of the structure. What *decreased* in number were jobs in the middle range that combined loose requirements with decent pay, such as blue-collar production jobs in large-scale manufacturing.

Table 6.2 shows these shifts in the occupational structure from 1972 to 1982. Unfortunately, although these are the best data currently available, 1972 represents a peak in the business cycle, while 1982 represents a trough. The result is to overstate the increase in the highest-level jobs, while somewhat understating the growth in lower-level employment.[1] Even with this bias, however, comparing white-collar jobs with service jobs shows comparable orders of increase, 36 percent and 33 percent, respectively.

Not all white-collar jobs are highly desirable; sales and clerical occupations, which tend to be low-paid and are more likely to be part-time and intermittent, actually account for a little over half of all white-collar jobs. By the same token, not all semi-skilled and low-skilled blue-collar jobs are better than white-collar jobs at similar levels, but many more of the former than of the latter have had some kind of mitigating shelter. This was particularly true of machine operatives in large-scale industry and other organized workers such as long-distance truck and bus drivers. While employment rose by 21 percent over the decade, these jobs showed an absolute decline. The drop might be slightly less if we had comparable data for 1984, but the decline in relative shares will not be reversed. This, then, is the disappearing middle, the cadre of displaced workers in the smokestack industries (Bluestone and Harrison, 1983; Freedman, 1982). Although these workers are not necessarily unemployed, there is little doubt that their traditional jobs are gone (Wachter and Wascher, 1983).

TABLE 6.2   Nonagricultural Workers by Occupation, 1972 and 1982
(in thousands)

	1972	1982	Percentage Change
Total employed	82153	99526	21.1
Managerial and professional	16108	23152	43.7
professional	8830	12555	42.2
executive, administrative, managerial	7278	10597	45.6
Technical, sales, administrative support	23619	30769	30.3
technical	1928	3013	56.3
sales	8566	11249	31.3
administrative, including clerical	13125	16507	25.8
Service occupations[a]	9391	12451	32.6
Private household	1440	1043	−27.6
Craft and repair	10347	11775	13.8
Operators and laborers	17384	16550	− 4.8
machine operators, assemblers	8600	7874	−8.4
transport and material moving	4143	4198	1.3
helpers and laborers	4641	4478	−3.5

SOURCE: Klein (1984: 14).
a. Excludes private household workers.

THE WEAKENING OF LABOR MARKET STRUCTURES

In terms of how the labor market operates, the displacement of blue-collar workers is only one incident. On a far larger scale, there has been a weakening of attachment and a loss of protections. More women are working and more jobs have the characteristics of traditional female-labeled jobs, which is to say that they are more likely to be part-time or intermittent, with briefer tenure and lower pay than the average for their requirements.

But if the increasing labor force participation of women speaks to this question in a general way, there is also more specific evidence of weakening structures. Audrey Freedman of the Conference Board was quoted as concluding that in the 1980s, unions disappeared as a major factor in wage setting (Sawyer, 1984). Whether or not the case should be put so strongly, there is little doubt about the trend.

A new era in labor relations was inaugurated by the breaking of the air controllers' strike in 1981. The temporary cutback in scheduled flights, followed by the cutthroat competition introduced by de-regulation, coincided with the severe cyclical downturn of the period to put the major airlines in economic jeopardy. One response was to take a tough line with the unions. Braniff, which suspended operations in 1982, resumed in 1984, but with a greatly reduced staff and redrawn pay scales and working conditions.

Continental Airlines took a pioneering step by declaring bankruptcy even though it was not out of cash, and then cutting its work force by almost two-thirds and reducing wages by half. The Supreme Court subsequently ruled that companies that file for reorganization under the bankruptcy laws can cancel a labor contract even before a bankruptcy judge has acted, thus giving healthy as well as ailing companies a major bargaining tool.

Both Western and Eastern Airlines gained concessions from workers in exchange for stock, but in an even greater departure from the past, American Airlines, with union agreement, introduced a two-tier wage structure, in which workers hired after a contract is negotiated receive lower wages than those already employed.

Within a few days of the beginning of a strike against the Greyhound Bus Company in 1983, 20,000 people applied for the strikers' jobs. After almost seven weeks, the workers voted to accept a 7.8 percent pay cut to get their jobs back. Meanwhile, the second largest long-distance bus company, which already had a lower pay scale, used the occasion of deregulation to divest itself of its small, unprofitable runs by spinning them off to independent operators, but keeping them in their system for scheduling purposes.

The rapid decline of copper mining has made union-busting relatively easy. Phelps Dodge, for example, defeated its union and resumed operations in Arizona, although at least 60 percent of the state's copper miners are permanently out of the industry. In other industries, operating with weaker unions or no unions does not guarantee success. In the beef sector of meatpacking, the union began to lose ground to nonunion producers as far back as the 1960s. In pork, many companies are being spun off as nonunion or low-wage operators by the conglomerates that acquired them in the 1970s. But wages, which account for 10 to 15 percent of production costs, are manifestly not the only problem. The union at Wilson accepted a 44-month wage freeze in December 1981, only to have the company declare bankruptcy in June 1983.

Even in cases where companies have not pushed their antiunion efforts to the limit, the effects of deregulation, and especially the entry of new firms, has tended to weaken wages, benefits, and working conditions. Two major examples are in trucking and telephone. Thousands of new nonunion truckers entered the long-haul business after the Motor Carrier Act of 1980 authorized the Interstate Commerce Commission to lower the barriers to entry. In June 1983, the number of carriers was about 40 percent higher than it had been three years earlier, creating overcapacity and deep rate discounting (Salpukas, 1983).

The deregulation of the telephone industry, culminating in the AT&T divestiture at the beginning of 1984, had a similar effect. Non-Bell carriers tend to be nonunion, and as they have taken away old accounts and created new ones, the industry is becoming both more competitive and less unionized. In addition, the interaction between technological change and deregulation is creating a new labor market picture with respect to the number as well as the type of workers required (Freedman, 1984).

Still another factor in the undermining of labor market structures is the regional shift of population and economic growth away from the traditional industrial centers. For example, construction unions are traditionally weaker in the Sunbelt than elsewhere. As the local economies of the Southwest have expanded, the growth in contract construction and special trades contractors has not added strength to the organized sector. Rather, new firms operate in the unstructured fashion of the homebuilding part of the industry.

Up to this point, all we have shown is that employment shelters in private industry have been weakened and that the jobs where shelters were strongest have steadily declined as a proportion of total employment. Another aspect is the shift in the relative size of class-of-worker categories. While there are fewer new opportunities for stable jobs at decent pay in either the private or the public sector, self-employment, the least stable type of work, has been increasing since the early 1970s.

SELF-EMPLOYMENT AND ENTREPRENEURSHIP

In 1948, a few years after the end of World War II, 12 percent of the nonagricultural work force was self-employed. The majority were small proprietors precariously in business for themselves. Already, however, there was a growing tendency for workers to seek stability in working for others, preferably for "the company"—the utility, the phone company, the dominant manufacturer of the locality. The economy was generally booming; American industry was busy rebuilding and retooling the war-ravaged world at the same time as it was being called on to satisfy the pent-up demand of the war years at home and the new demand created by the postwar population explosion. It was a time when security of employment seemed to become not only a goal, but also a norm, when eminent scholars worried about the "new industrial feudalism" that might impede labor market mobility and efficiency (Ross, 1958). These trends, however, turned out to be reversible; one concomitant is that larger numbers of people are once again going into business for themselves.

At the end of 1983, the proportion of the nonagricultural labor force that was self-employed reached 8.6 percent, the highest since 1966. This figure somewhat understates the case because in addition to the 7.8 million who reported that they were self-employed in their main work activity, about 5 million people had two or more jobs, and some of these second jobs were undoubtedly in the free-lance category.[2] The size of the increase in self-employment in the 1980s is in some measure a tribute to the severity of the economic downturn, but it is also related to the structural changes outlined above. Both factors are implicated, for example, in construction where self-employment increased 5.5 percent per year during the 1970s, apparently as a result of the relative increase in residential activities involving repair, maintenance, and renovation, which are typically performed by small firms. Further increases in the 1980s were related to the traditional counter-cyclical nature of self-employment (Tschetter and Lukasiewicz, 1983). That is to say, when times are bad, people calculate that they have little to lose by taking the risk of going into business for themselves.

Interestingly enough, the growth of self-employment and small business has been widely hailed, both as a return of the American entrepreneurial spirit and as a prime generator of jobs. One view is that clever people, who cannot even be sure of security if they work for large companies, are in any case restless in the grip of bureaucratic management (Wayne, 1984). For every spectacular success, however, there are a large number of failures: The majority of new business firms do not survive four years (Armington and Odle, 1982). In 1983, the first year of the recovery, business failures went up 24 percent, and showed no signs of bottoming out in 1984, as they usually do in the second year of a recovery (Cuff, 1984).

As to employment growth, the popular idea that small businesses account for most new jobs in today's economy turns out to be based on confusing small establishments with small firms. If the data are examined by size of company rather than by how many people work in a given location, the small-business (firms with fewer than 100 employees) contribution to net job growth is about proportional to its share of the American labor force. What is more significant is the fact that small businesses have a much higher share of growth in slow-growing industries and in lagging regions because they tend "to enter and expand into relatively weak areas of production (those without large-business competition)" (Armington and Odle, 1982). What this means is that those with a taste for entrepreneurship may find a place in the interstices of the economy that are outside the purview of large-scale capital enterprise, even though these marginal firms do not generate a great deal of employment.

The changed conditions in the labor market outlined above have differential effects on employment prospects, depending on both the information and access available to particular groups and the pathways of skill acquisition and training typical of the occupation in question. Immigrants and the native population do not compete on a one-to-one basis, even at the same level of skill. Native workers have the advantage in jobs where labor market structures are strong, while immigrants have their best shot at jobs in the more loosely organized sectors of the economy. As market protections weaken, newcomers from abroad have, by so much, the better opportunity.

The issue of competition has been most in the forefront as it affects native Blacks. Those who view immigrants and Blacks as unskilled labor competing for low-wage jobs miss the point. Blacks who succeed in establishing themselves do so by completing professional courses of training that provide them with credentials for jobs at the upper reaches of the occupational structure, or by access to public service employment where civil service rules provide protection, or by access to unionized employment. Only the first of these routes, professionalism, is expanding, and as opportunities narrow for those with no special qualifications, Black workers most in need of labor market intermediation are precisely the ones with the least access.

Studies in Los Angeles that show little effect on Black employment in the presence of an increase in the foreign-born work force (Muller, 1984) are borne out by data from other areas (Bailey, 1984). The statistical findings alone give the appearance of Black-immigrant complementarity, but underneath the surface, one can discern divergent employment processes that channel groups of workers into particular sectors of the labor market.

The question of access is most critical for entry workers. The younger and less experienced the applicant, the more important is the intervention of family and friends. A high-school dropout hired by a relative to help out in a store or a gas station has a job created just for him or her. High-school graduates working as data-entry clerks in a back office become a source of information for their social circles about openings in a high turnover occupation. The prospect of passing a civil service examination, or even to find out where and how to sit for it, is enhanced by acquaintance with postal clerks and police officers.

Once identified with an occupation, it is the occupational network that for better or worse becomes the most important factor in labor market information. Professional associations provide both formal and informal knowledge about the prospects in the job market. Large companies make it possible to bid for promotion or transfer. Whether the work involves typing, turpentine collecting, or computer programming, workers in occupations with well-defined content find it relatively easy to learn what others are earning and where there may be openings.

In all these examples, job seekers get to know what particular firms require. There are many variations in detail, but all employers like to have reassuring references. A positive employment history and attestation to the applicant's character go a long way in resolving doubts about reliability, trustworthiness or willingness to cooperate. The problems that this raises for young Blacks emerge from a study by Henry Jay Becker (1979: 5):

> An effective reference comes from someone whose evaluations would be held in high regard by the potential employer—for example, other employers and business associates, or the employer's most trusted employees. In contrast, the sponsors that most young black job-seekers are able to provide are their own friends and relatives, nearly all of whom are likely to be black and in most cases unlikely to impress an employer.

Becker's point is not that Blacks are unwilling to recommend their family members and friends, but rather that many are not well-placed to do so. Kinship is just as important a source of support among minorities as among other groups (Stack, 1974), but where status achievement is limited, so too is the help that is available.

Immigrants, in the meantime, are finding their way in the interstices of the economy, in those small enterprises that arise at the margins and particularly in businesses established by their compatriots. It is generally acknowledged that self-employment and petty entrepreneurship are critical for the settlement process. What is less obvious, perhaps, is that skill acquisition flourishes in these settings. It turns out that the jobs being created in our transformed economy do not require structured internal labor market arrangements. On-the-job training still takes place, but not necessarily in formal settings or in connection with hierarchical job ladders.

IMMIGRANT ROLES IN URBAN LABOR MARKETS

Gross data on the occupational distribution of the foreign-born lend themselves to a variety of interpretations. Thomas Muller (1984: 8), for example, finds it unsurprising that Mexicans and other immigrants are streaming into Los Angeles. The reason lies in the city's concentration of manufacturing industries:

> Just as the Irish dominated the factories in Boston, Slavic immigrants filled the mills in Pittsburgh and Cleveland, and Jewish immigrants were the majority group in the garment industry in New York, Hispanics now provide the low-wage labor in Los Angeles industry.

But newcomers, Hispanics and others, are also streaming into New York City where manufacturing employment has steadily declined for decades. In both cities, the foreign-born account for a disproportionate share of low-wage workers in factories, but that does not imply that the demand for their labor is an independent phenomenon. Their presence contributes to the maintenance, or even the expansion, of low-wage activities in a region, but the same jobs would exist elsewhere if not in New York or Los Angeles. The market is international, and from an employer's point of view, it is only a matter of convenience (with transportation costs and tax advantages weighed in the balance) whether electronic components are assembled on one side of the Mexican border or the other. In most marginal shops, immigrants themselves play an entrepreneurial role, thus creating the jobs that absorb the labor of those who continue to arrive.

In Los Angeles, in New York, and indeed in the country as a whole, the post-1965 immigration is made up of a variety of ethnic groups who bring with them a variety of skills and abilities. What they have in common, as we have already pointed out, is a well-developed network of family and kinfolk. Three factors sort them into labor market sectors— how much English they command, whether they have developed skills that are acceptable in the United States, and in what industries their compatriots are already established.

The less their control of the English language, the more closely immigrants are bound to their own ethnic groups. In Miami, where Cubans have developed an enclave that amounts to an economy within the economy (Wilson and Portes, 1980) this problem is minimal; in fact, it is the rest of the population that is disadvantaged without a knowledge of Spanish. Usually, however, problems with English constitute a severe restraint, as for example, among those well-educated Koreans in New York City who have settled for running vegetable markets (Kim, 1981).

The skill question is far more complex. Immigrants trained abroad are often unable to practice their professions in the United States because of licensure problems. Many who came here as students, on the other hand, find ways to adjust their status and become resident aliens after their American training is completed. By 1980, 15 percent to 20 percent of the foreign-born work force in New York City and in Los Angeles (among those who emigrated to the U.S. after 1965) were in professional, managerial, or technical occupations, compared to 28 percent of the labor force as a whole.

In skilled blue-collar jobs, new immigrants were actually over-represented in both cities. Foreign-trained workers may be the major source of supply in certain trades, as Israelis and European Jews are in diamond-cutting and setting. In others, skilled immigrants are welcome as a way of avoiding the cost and the trouble associated with training. Every immigrant construction worker who arrives with developed skills obviates the need for an apprenticeship or equivalent experience for a native worker.

Skill acquisition also takes place on the job. Those large U.S. firms that have well-developed training schemes for nonmanagerial workers organize matters in such a way as to obviate the need for accommodating teachers. Situations are structured to ensure that workers will eventually learn what is required of them even if no seasoned mentor makes an effort to train them. Among small firms, however, learning is perforce informal. Familiarity with coworkers' values and personalities, to say nothing of their language, is crucial in gaining their friendship and support, not only for emotional comfort, but also for developing productive and organizational skills. Informal training is largely a function of the amount of time workers wish to spend together and how much energy they put into the relationship. Work breaks, which provide time for skill acquisition, are more likely to be shared among people of similar backgrounds. If teamwork is required, workers must learn to interpret each others' cues and anticipate movements, coordination that is hard to achieve when there is a high degree of conflict (Kornblum, 1974). Valuable learning occurs when workers fill in for absentees. The sponsorship of a coworker may be helpful in convincing a supervisor to allow a new worker to try out interesting and challenging tasks. When these conditions are met, skill acquisition can take place in settings usually associated only with menial work, such as restaurant kitchens where immigrant restaurateurs create a skilled labor force in the course of ordinary business operations (Bailey, 1984).

The small businesses of immigrants play a key role for newcomers, regardless of the possibility of skill acquisition. The typical small shops, restaurants, and factories established to meet specialized market demands do not create large numbers of jobs; in fact, many of them are run by families, by mom and pop alone, or together with a few relatives. What they do provide is at least a way-station and at most a long-term job placement for the most recent arrivals.

Again, we should emphasize the dual role of kinship, first in fostering the immigration chain and, subsequently, in finding jobs for the newcomers. Chinese from Hong Kong join their families in the United States; the men are absorbed in the restaurant industry, while the women find a place in the garment shops. These businesses, established by earlier immigrants, do not always succeed, but others arise to replace them. Chinese cuisine is an exceptionally attractive and inexpensive alternative to fast food, on the one hand, and to the high-priced full-service restaurant, on the other. Thus, although Chinese restaurants compete with other restaurants, to a marked extent they create their own demand.

The garment shops play a different economic role. They permit relatively inexpensive clothing to be produced in New York or Los Angeles at costs that are competitive to contracting-out abroad and at a speed that fulfills the demand of the typical spot markets generated in the industry (Waldinger, 1982). The apparel industry has provided work for immigrant women for generations, but in employment terms, causation runs the other way just as it does in restaurants: the labor supply provided by immigrant women maintains the industry.[3]

Other loosely structured situations also offer opportunity. In the mid-1970s, operating costs in the New York City taxi industry made large fleets less attractive. Since profits were small, it paid to sell taxi permits ("medallions") to individuals and minifleet owners who were willing to exploit themselves in exchange for economic independence. Small collectives with radio dispatching in dozens of languages now operate in the city, together with flourishing car service companies that respond only to telephone calls. Israelis are particularly active in this latter business, helped along by the existence of a parochial Jewish clientele in Brooklyn that prefers to deal with other Jews (Freedman and Korazim, 1983).

Such examples can be reproduced in any city with a large cohort of foreign-born residents. The very terms of their immigration bind them together in ways that enhance the possibility of finding a job and making a living, even if the job is onerous and the living is marginal.

THE LABOR MARKET DISADVANTAGES
OF BLACK WORKERS

As we have already remarked, there is little evidence of direct labor-market competition between native Blacks and new immigrants, but insofar as the latter seem to create new, if perilous, footholds in the urban economies of America, the question often arises of why Blacks do not, in effect, go and do likewise when more structured opportunities are not available to them. To address that question, we need first to look briefly at the historical record.

In the 1980s, the combined effects of economic downturn, the undermining of protective structures in the labor market, and the weakening of public programs in aid of the poor put a halt to Black social progress. This is not the first time that Black gains have been reversed.

After the Civil War, the best thing that could have happened to black workers of the United States would have been a fair opportunity to contribute to satisfying the great demand for labour in the rapidly growing cities of the North and West . . . [to have] earned their place . . . in an expanding economy which permitted a continually changing equilibrium between a succession of very different ethnic groups [Thomas, 1961: 330].

But Blacks were not just another ethnic group. It took 100 years to achieve equal legal rights, and in all that time and beyond, color itself retained ascriptive importance. For most of this period, a major disability suffered by Blacks was occupational segregation, which worked against their taking their place in the ethnic succession, especially in the critical period of European immigration around the turn of the century (Bodnar, 1976).

Over the years, however, Blacks did make certain gains. Immediately after the Civil War, former slaves began to accumulate wealth in the form of land and farm capital. At the peak, these holdings were miniscule, but by 1915, even this minitrend toward property holding came to an end (Higgs, 1982). In this period, the vast majority of Blacks lived in the South under conditions that approached peonage and that failed entirely to prepare them for the great migration to come. For the minority who had been born in the North, opportunities actually diminished and residential segregation became more severe.

The educational advantage of Whites over Blacks had narrowed, and northern Blacks had actually advanced slightly more rapidly than the second-generation "new Europeans" from the migration waves of the late nineteenth century. By the 1920s, however, the relative rates were reversed, and finally there was a massive upward surge in the educational gains of Europeans over Blacks (Lieberson, 1980).

This reversal in educational achievement was due to the hardening of white attitudes. After World War I, northern schools adopted segregation schemes and weakened the curriculum offerings in predominantly Black schools and classes. But in the labor market, the gap widened earlier. By 1900, immigrants were already moving into intermediate blue-collar jobs and skilled trades where Blacks had miniscule representation (Bodnar, 1976). By the 1920s, a distinct upgrading was apparent in the position of White immigrants relative to Blacks in northern factories. With the returns to education less for Blacks than for Whites, it is hardly surprising that among the 1925-1935 birth cohort, there was a sharp increase in education among "New Europeans" and a sharp decline among Blacks (Lieberson, 1980).

The great Black migration from the south began in the 1920s with the collapse of the cotton economy, and the situation of those who remained became more desperate during the Depression. Even then, however, Blacks could only buy land if the sale was initiated by a White. Tenancy and sharecropping meant dependency; the failure of the states to support public education for Black children meant ignorance (Raper,

1974). Blacks were the reserve army of agriculture in the country and of menial service in the city. Black professionals served only Black clients, and Black-owned businesses existed only in Black areas and in activities that Whites found unprofitable. Such a society did not even require a concept of racism; the consensus found Blacks to be inferior and there the matter rested for decades.

The situation in the North was only marginally better. As the number of Blacks arriving from the South increased, so did spatial isolation. Originally new European immigrants had been more segregated, but already between 1910 and 1920, the situation was reversed "because whites in each city were attempting to maintain the degree of isolation from blacks that existed before the new flows from the South started" (Lieberson, 1980: 291).

It is critical to keep in mind that Blacks were never immigrants; among other implications of this fact, they were cut off from their past and had no successful forerunners to smooth their way in the new land. Ironically, just at the time when their improvements in status were being lost in the North, those who were arriving from the South were more like immigrants than they would ever be again. Their aspirations were rooted in comparisons with the South, and they were probably as willing as immigrants from abroad to suffer the pains of working under industrial conditions that, although harsh, were more favorable than what they had left behind in the countryside. They found jobs but not an equal place, and it was this early skill deprivation that cost them the battle for succession that immigrants from abroad would win.

After several generations, the children of European immigrants established themselves, first through the strength of their unions, and later through moving out of the factories and mills. Blacks, meanwhile, continued to suffer severe discrimination. *Brown v. Board of Education,* the case that finally ended legally imposed school segregation, was decided only thirty years ago. In a kind of parallel to the postbellum period, the new freedoms gained through that decision and the subsequent victories of the civil rights movement were limited in their effects.

An important point here is that Blacks have had to put their main reliance on government for protection and for equalizing opportunity. It has been in periods of strong federal support that they achieved significant progress, and it is education that has had the most enduring effect. That is because the formal channels provided by the educational system have been easier to monitor than the informal networks that pervade American life. Affirmative action, while having a positive effect on higher education, had a more significant (albeit limited) impact in overcoming employment barriers because it acted to establish individuals in places where they could create new networks.

Public intervention, for example, helped to open up the building trades. Blacks had always had a large share of the laborers' jobs in the industry, but just as they began to gain a weak foothold in the skilled trades, construction declined as a source of employment. As in other declining industries, nonunion firms with less structured work rules began to grow in proportion to the union firms where Blacks had made the most inroads. Interestingly enough, Black-owned firms in New York City are often headed by business school graduates, rather than by the usual sort of self-made contractor, and it is the preference accorded to them in awarding contracts for public projects that seems to account for their survival (Gallo, 1983).

BLACKS AND IMMIGRANTS

Immigrants have found it easier than Blacks to acquire informal training on the job. And it is the strength of their links to such opportunities that is both a cause and an effect of their higher incidence of self-employment and entrepreneurship. Compared to Blacks, immigrants have found clienteles more willing to deal with them and encountered fewer difficulties in raising capital.

Asian immigrants, who are among the most entrepreneurial of the recent immigrant wave, have accumulated capital in several ways. Some, like Indians who received severance pay after a period as *gastarbeiter* in Europe, or like Koreans who cashed in their assets before emigrating, come to the United States with enough funds to get started in a marginal enterprise. Others raise money after they arrive by working for compatriots while learning a business. Still others are able to get loans, either informally from friends or formally through the banking system. Few of these means are available to Blacks, and without them, they are unlikely to compete as petty entrepreneurs.

It is, in general, rather late in our history to expect Blacks to behave like immigrants. Their problems and prospects are complicated by issues rooted in the past. After emancipation, they were accorded grudging acceptance by White society as long as they stayed "in their place"; subsequent efforts to move out were met with residential flight and, finally, resistance to school integration. Immigrants, who endured much, did not experience this measure of exclusion and suspicion on a continuing basis. On the other hand, immigration was not the cause of the special burdens borne by Blacks. In the parallel developments surrounding the industrialization of America, immigrants filled certain roles that were denied to Blacks, and, as a consequence, the children of immigrants were able to go further and do better in the presence of economic expansion.

CONCLUSION

At the end of the 1960s, the U.S. labor market entered a new period in which post-World War II trends began to be reversed. Among the important transformations were a decline in the birthrate (which would have its long-run effects in the 1990s), an increase in the labor force participation of women, a shift in industries that caused declines in employment in manufacturing and growth in employment in services, and a weakening of the ties that bind workers and employers together.

In the middle 1970s, and again in the 1980s, deep recessions interacting with technological advances and new competition from abroad served to weaken the position of workers vis-à-vis employers, and while those who remained employed in selected industries continued to enjoy certain protections and fringe benefits, the situation of many workers (not including the top strata among professionals and managers) became far less certain than it had been in the past. Incomes grow more disparate, but at the middle and the bottom of the earnings distribution, conditions are converging, with more and more jobs being characterized by uncertainty.

Real earnings have been steady in the decade from 1975 to 1984. Insofar as families can satisfy their aspirations, they do so by having more than one income. Up to approximately a $70,000 annual income, there is a strong correlation between income level and the presence of two earners (usually husband and wife; Rose, 1983: tables VI-VIII). Single heads of household with children have not kept up; they constitute the bulk of those who live in poverty.

After the 1965 revision of the law, a new wave of immigrants (joined by refugees from time to time) began to arrive, mostly from Latin America, the Caribbean, and Asia. These newcomers were mostly poor, at least upon their immediate arrival, but self-selected in two respects. Traditionally, it is the more able who take the step of emigrating, but perhaps more important in the new U.S. situation, reuniting families became a major goal of immigration legislation. By definition, therefore, newcomers had family support. This often involved employment in an activity being carried on by the family itself or by other kin or countrymen. Networks of information and access tend to be ethnically limited among newcomers, if for no other reason than language and culture. The skills that immigrants bring can only be released in a setting where their language can be understood. In exchange, the workers are protected by a paternalism that provides not only a breathing spell, but often an opportunity to learn further skills.

Because of their special access, immigrants do not directly threaten the jobs of native minority workers. The competition is mediated by intervening factors. In the restaurant industry, for example, immigrant-owned restaurants tend to lessen the market share of fast food outlets, which typically provide employment for such disadvantaged groups as Black teenagers. Small nonunion construction firms, which flourish in the renovation sector, have an advantage if they can find already trained craft workers among newcomers from abroad, but someone in the firm has to be able to communicate with and supervise them. Small retail businesses are risky enterprises, but they are more likely to flourish if family members are willing to exploit themselves.

These are not the typical pathways of native Blacks. Their continuing segregation, inability to find White customers, difficulty in raising capital, and general lack of experience all militate against successful entrepreneurship. Historically, Blacks have made the most progress when government has responded to their needs as a matter of equity. Public employment, government-sponsored training programs, educational assistance, and affirmative action have all played a role in advancing the economic position of the minority population. When such aid has been withdrawn or weakened, their progress slows down.

Meanwhile, as the birthrate continues to decline, the pool of new workers will not grow as quickly as in the past. This demographic shift is not in itself a cause for optimism. If all groups had equal opportunity, then we might look forward to less job competition, but in fact, uneducated Blacks seem to be permanently at the end of the queue, if not altogether outside it. Some Hispanic groups also remain disproportionately disadvantaged, especially when the second and succeeding generations find themselves unable to leave their segregated neighborhoods and where schooling fails to provide a route for upward mobility. The constraints on opportunity for less educated workers affect all groups, but for those who remain identifiable by reason of color or an incomplete transition to standard spoken English, the prospects are even dimmer.

One thing seems clear—the children of immigrants, the first generation born here, will not be as willing as their parents to "kill themselves" working for survival. No longer immigrants, they will share the bitter and the sweet of aspirations and achievements with the rest of the native population. Insofar as ascriptive qualities bar their progress, and residential segregation turns them in upon themselves, by so much will they swell the ranks of the poor. Immigrants are not immune from the problems of American society. The seeming dispensation that makes it possible for them to survive in the beginning may or may not prove beneficial later on. Among high achievers, we know that weak ties— those in the profession or the workplace—are more important to advancement than the strong ties of family and kin that sustain newcomers from abroad (Granovetter, 1974). The ethnic network contributes most when a significant number of one's group succeed and provide not only the road map, but also the vehicles, for those who come after. How the transition is made to a different style and a different type of network may turn out, in the event, to be the crucial difference in how groups shape up statistically as their American roots deepen.

NOTES

1. The changes in occupational definitions for the 1980 census obviate the possibility of direct comparisons in the usual way. The Bureau of Labor Statistics, however, did one sample study in which 1972 and 1982 occupational data were both coded using 1982 definitions (Klein, 1984). In the 1983-1984 recovery, employment in low-level jobs increased, but these jobs were mainly in services rather than in the higher-paid manufacturing sectors.

2. The idea that large numbers of employed workers are unaccounted for because they are in the "underground economy" turns out to be an attractive piece of mythology. The Bureau of Labor Statistics has marshaled impressive evidence that, for example, some of the self-employed may be tax evaders, but that does not mean that they are excluded from the labor force count (McDonald, 1984).

3. In capital-intensive industries such as automobile manufacturing, Japan can compete by investing in the industry in the United States (in the same fashion as the United States historically did in Europe) if it is blocked from selling in the United States market. In a labor-intensive industry such as apparel, there is no similar incentive. Instead, the activity is maintained by the presence of the same sort of labor supply as supports it in less developed countries.

7

Political Economy and the Social Control of Ethnic Crime

ELEANOR M. MILLER
LYNNE H. KLEINMAN

IDEOLOGICAL CONTEXT

The nature of the interaction between ethnic groups in the United States and the criminal justice system has in some measure been determined by attitudes and convictions held by those who have been responsible for or influential in formulating criminal justice policy. These attitudes and convictions have clearly emerged from an acceptance by some public policymakers of the notions that ethnic groups are somehow "inferior" and, more important, have a greater tendency toward criminal involvement than does the general population. Such notions about the characteristics of ethnic groups have their roots in identifiable political and intellectual movements in the history of the United States. It is to an examination of these roots that we initially turn our attention.

The belief in Anglo-Saxon superiority provides one key explanation of the way in which American attitudes toward ethnic minorities developed. Our expansionist impulses, expressed in such concepts as Manifest Destiny in the nineteenth century, were strongly tied to convictions about White racial superiority over Native Americans and over any other racial or ethnic minority that attempted to assert prior claims to the land. Similarly, the imperialist impulse, which climaxed with the Spanish-American War of 1898, was based upon a belief in Anglo-Saxon racial superiority over all the world's "backward peoples" (Hofstadter, 1944: ch. 9). Higham points out that these impulses laid the groundwork for the twentieth-century development of racism among northern intellectuals, which was directed against all groups of non-Anglo-Saxon background. It accounts as well for the popular hatreds that emerged especially in the South and West directed against specific racial and ethnic minorities, chiefly Blacks and Asians (Higham, 1967: 170).

A second key thread in the development of attitudes toward ethnic minorities emerged from the eugenics movement that lasted from 1900 to 1915. This movement accepted the older Social Darwinist concepts of the upper, privileged classes as "fit" and the lower, downtrodden classes as "unfit." It advocated improvement of the human race "by breeding from the best and restricting the offspring of the worst" (Higham, 1967: 150). Pauperism, disease, and immorality (including criminality) were viewed as inherited characteristics, and the large urban centers, housing ethnic minorities in the form of immigrant populations, appeared to contain ever-increasing numbers of "the diseased, the deficient, and the demented" (Hofstadter, 1944: 162). Because the interest in eugenics brought hitherto unrecognized cases of diseased and defective families to the surface, the number of known cases was misinterpreted as a real increase directly due to immigration. For our purposes, it is important to realize that the eugenics movement clearly affected the development of attitudes toward ethnic minorities, and that these attitudes were expressed (at the beginning of the twentieth century) in terms of immigration restriction. The essence of the matter is this: The work of the eugenicists "made virtually every symptom of social disorganization look like an inherited trait", and it was axiomatic that this was not subject to modification by environmental factors (Higham, 1967: 151). Thus emerged a rationale for excluding any ethnic group whose members exhibited difficulty in assimilating into American life (the range of possible "difficulties" was wide: language problems, problems arising from the practice of peculiar customs, physical or mental disability, criminality, and so on).

The third key element in the development of attitudes toward ethnic minorities was a brand of nativism that emerged concurrently with the eugenics movement. This was influenced both by that movement and by the older beliefs in Anglo-Saxon superiority. From the turn of the century until the outbreak of World War I in 1914, nativists focused their attention on the shift in national backgrounds of immigrant groups entering the United States. No longer was the bulk of immigration coming from northern and western Europe (chiefly the British Isles, Scandinavia, Germany, and the Low Countries). The immigration of the early twentieth century (which was three and a half times the size of the nineteenth-century immigration) was coming, instead, from southern and eastern Europe. Nativists now were moved to make a distinction between the "old" and "new" immigration, the former being considered much more desirable because of its clear Anglo-Saxon lineage and demonstrated adaptability to American ways. The new immigration, by contrast, was perceived as causing nothing but problems for the United States. This was exemplified, for example, in the great attention given to criminal behavior among Italians:

> It was universally believed that serious offenses were rapidly increasing— as they probably were. Lax American law enforcement was attracting to the United States a considerable number of Sicily's bandits; here, through blackmail and murder, they levied tribute on their intimidated countrymen more successfully than they had at home. By 1909 (and this is the critical factor from the point of view of attitude-development), when a combined drive of American authorities and Italian community leaders began to reduce these activities, the image of a mysterious Black Hand Society, extending from Italy into every large American city, was fixed in the public imagination [Higham, 1967: 160].

Thus the ethnic minorities who constituted the new immigration became associated in the American mind with undesirable behavior stemming from undesirable character traits. And the eugenicists warned that these character traits were products of heredity, unalterable by environmental influences. Consequently, the early twentieth century saw the development of a strong immigration restriction movement, geared specifically to halting the flow of undesirable ethnic groups. Because the political expediency of institutionalizing a policy of immigration restriction was highly questionable, a delaying tactic, in the form of a legally mandated investigation of the issue, was employed. On February 20, 1907, Congress enacted legislation establishing an Immigration Commission that consisted of three senators (chosen by the president of the Senate), three representatives (chosen by the Speaker of the House), and three citizens (chosen by the president). The Immigration Commission was chaired by Senator William P. Dillingham of Vermont. In 1911 it published a 42-volume report of its findings that is highly significant as an indicator of the attitudes of public policymakers toward ethnic minorities.

In a critical evaluation of the Dillingham Commission Report, Handlin has observed that the commission began by taking for granted the conclusions it wanted to prove, namely, that the old and new immigrations were different in character and the latter was less capable of being "Americanized." The preparation of the entire report, Handlin (1957: 81-82) says, was conditioned by this assumption, and thus there was no effort made to trace the development of various problems by comparing earlier and later conditions; no use made of available census data; no use made of information collected by other government agencies; no account taken of the discrepancies between the experiences of different immigrant groups based on duration of settlement.

Despite this fact, the commission drew up a composite picture of "races and nationalities . . . exhibiting clearly defined criminal characteristics" (U.S. Immigration Commission, 1911, Vol. 36: 2). Italians were most likely to be responsible for crimes of personal violence. Native citizens were only slightly more likely than Jews to commit crimes against property such as larceny, burglary, and receiving stolen goods. Greeks, Italians, and Jews were all frequent violators of city ordinances regulating peddling. The Irish were singled out for their drunkenness and vagrancy. And the French and Jews were particularly likely to be among those arrested for prostitution (1911: 16-20).

Handlin illustrates the shortcomings of the entire investigation in an analysis of the commission's report on crime. This report stated that the incidence of crime among the foreign-born was proportionately higher than among the native population, completely ignoring the U.S. Census report on prisoners that indicated that immigration had not increased the volume of crime significantly and that the percentage of immigrants among prisoners had actually fallen. Handlin's point, and the one that most concerns us here, is that the Dillingham Commission ignored the findings of other responsible investigators that did not support the conclusion it had decided upon in advance. Indeed, the commission very deliberately collected its own figures on incidence of crime from urban areas with high immigrant-density populations (New York, Chicago, and Boston), which clearly were unrepresentative of the country as a whole, and organized its data to show that immigration had changed the character of crime in the United States. Specifically, immigration was said to have increased violent crime and such offenses against public decency as drunkenness, vagrancy, and prostitution. Handlin's close examination of the data reveals that although the committee *said* it was comparing incidence of crime among the foreign-born to incidence among the native population, it was *actually* comparing the incidence of a particular kind of crime committed by a group to the total incidence of crime in that group. In short, there was no real comparison made between ethnic groups and the native population (Handlin, 1957: 99-102).

The Dillingham Commission did not attempt to explain the correlation between crime and ethnicity. It can perhaps be assumed that theories of genetic inferiority were so much a part of the folk wisdom of the time that the etiology of such a relationship did not need to be explored. That the observed correlation might be a by-product of poverty simply was not considered. To consider environmental factors was to allow for the fact that they could change, and thus the impugned behavior might also change. It is clear, however, that "foreignness" was an evil in and of itself. One of its many problematic manifestations was criminal propensity. Thus, for the commission, the relationship between crime and ethnicity was self-evident, its explanation simply assumed and its remedy obvious: Restrict the immigration of those "races" prone to crime.

MATERIAL CONDITIONS:
THE NINETEENTH CENTURY

The industrialization of America began in the 1830s. The international preeminence of the United States was primarily established after the Civil War, but it was especially during the 1840s and 1850s that per capita productivity really took off. During that time, it increased at a rate somewhat greater than that of the long-range trend for the entire nineteenth century (Ward, 1971: 23).

Population growth and exceedingly rapid urbanization went hand in hand with industrialization. Between 1800 and 1850, the nation's population increased by a factor of 3.37 (from 5.3 million to 23.1 million). During that same time urban population grew by a factor of 10 (from 320,000 to 3.5 million). Between 1850 and the end of the century, the population expanded by a factor of 2.8 (from 23.1 million to 75.9 million), and the urban population grew by a factor of 7.51 (from 3.5 million to 30.1 million) (Larson, 1984: 102).

There is much evidence that suggests that during this period, but particularly during the second half of the nineteenth century, the crime rate and fear of crime on the part of the general population grew apace, especially in cities (Silberman, 1978: 22). With the conclusion of the Civil War, the exploitation of the nation's natural resources was undertaken as never before. In the three decades between 1870 and 1900, nearly 12 million immigrants came to the United States to fuel economic expansion. This is more than one and one-half times as many as came during the preceding fifty years. The majority settled in cities (Larson, 1984: 104). Steinberg asserts that "by the end of the century, the connection between immigration and crime became something of a national obsession" (1981: 116).

The conditions of life in most urban immigrant neighborhoods were dismal. By 1860, nearly 90 percent of New York City's paupers were immigrants (Larson, 1984: 104). Over a span of eight years during the 1880s, Jacob Riis (Cordasco, 1968: 86) is reported to have said that

> 135,595 families in New York were registered as asking for or receiving charity.... These facts tell a terrible story ... in a population of a million and a half, very nearly, if not quite, half a million persons were driven, or chose, to beg for food, or to accept it in charity.

It is obvious that conditions that spawn this level of want also spawn alcohol abuse and crime. Glaab and Brown (1967: 94) describe the resultant deviance:

> Although the immigrant in poorer areas of the cities was often arrested for inconsequential crimes that might be ignored elsewhere, there is little doubt that a life of desperation and the ready availability of alcohol contributed to a high rate of serious crime among immigrants, particularly among the Irish. In New York in 1859, for example, only 23 percent of the people arrested were native Americans and 55 percent were born in Ireland. Most of the persons committed to the city prison in New York City during the nine year period from 1850 to 1858 were immigrants, and seven-eighths of the total were recorded as "intemperate." In 1860 when just about over half of the population of the city was foreign-born, about 80 percent of the 58,067 convicted of crime during the year ending July 1 were born in Europe.

Ironically, despite living conditions that were both squalid and mean for the majority of nineteenth-century urban dwellers, the data available for cities such as Boston, Chicago, and Buffalo indicate a steady decline in serious crime from the 1830s to the 1930s even though the overall crime rate steadily increases (Larson, 1984: 105-106).

Glaab and Brown (1967) suggest an explanation for these trends. It appears that immigrants are increasingly being arrested for infractions that would be ignored in others, particularly intoxication. Powerful segments of the native and old immigrants become, at this time, increasingly intolerant of deviant behavior. Roger Lane argues that violent crime now brought far more severe retribution. He attributes the increased concern with law and order to the regimentation of life demanded by industrial employment (1969: 451):

The factory demanded regularity of behavior, a life governed by obedience to the rhythms of clock and calendar, the demands of foreman and supervisor.

But, more broadly, it is very clear that the initial establishment of police forces in all the major cities in the United States in the nineteenth century was directly or indirectly motivated by a desire to control the largely immigrant masses who fueled capitalist expansion. The initial concern, then, was actually one of *social control, not crime control* (Eitzen and Timmer, 1984: 373).

Lundman (as cited in Eitzen and Timmer, 1984: 372) has tied the rise of the modern police establishment directly to elite interests:

The . . . factor responsible for the formal organization of municipal police was elite interest and involvement. Persons of power, wealth and prestige overcame the historic reluctance surrounding the police idea. Their motivation was purely economic—elites used their influence to create police who would protect and promote their vested interests.

Spitzer and Scull (1977) define these "elite interests" as those benefiting from a developing capitalist economy. They argue that the private, profit-oriented system for crime control used by the feudal aristocracy in England was simply inadequate for the social control needs of the rising American bourgeoisie. They contend, furthermore, that the market economy of the time required a stable and orderly environment, and that the "costs" of achieving such an environment were seen, ideally, by the bourgeoisie as being borne by the public sector rather than by them themselves. In short, the time had come when it made sense to pass along these costs to society in the form of socialized social control. Similarly, it soon became apparent that the use of "company goons" to police workers' lives and repress workers' organizations would ultimately lead to "crises of legitimacy" if administered under the aegis of individual capitalists. The repression would have to become more and more "public" as the boundaries between capitalist domains and the private sector began to dissolve. Thus, along with economic motives, there were important political and ideological incentives for the shift from private to public policing (Spitzer and Scull, 1977; Spitzer, 1981).

The result was that police forces were established in major urban areas at this time. In general the precipitating events were major civil disturbances in the cities in question. From the 1830s through the 1850s a number of confrontations occurred between residents of immigrant slums and native or old-immigrant city dwellers. Some historians argue that this was perhaps "the era of the greatest urban violence that American has ever experienced" (Brown, 1968: 50). It is estimated that during these years in Baltimore, Philadelphia, New York, and Boston there were all told 35 major riots. There were labor riots, election riots, draft riots, antiabolition riots, anti-Black riots, and anti-Catholic riots (Larson, 1984: 106).

The precise nature and etiology of these riots are unclear. Some were mass responses to perceived threats to old immigrant livelihood from immigrant newcomers. Others appear to have been instigated by those who sought to legitimate their efforts to institute a formal public policing institution. The objects of repression, those perceived as the threats in the majority of these uprising, were new immigrants, particularly the Irish. An excerpt from the Boston Debates of 1863 illustrates the view of the city's White Protestant elite in this regard. Charles M. Ellis, Esq. (as quoted in Eitzen and Timmer, 1984: 373), speaking in favor of the Metropolitan Police Bills argued:

> In cities there are large bodies who are amongst us, but not of use, whose sentiments and principles do not harmonize with those of the State; who have influence, though they have not rightful influence; or, if citizens, have not the same influence as the great body of citizens—a foreign, floating population—those controlled by extraneous, alien influences, trade, criminal or other. You have amongst you those elements of society called . . . the dangerous classes.

There is evidence as well that efforts to professionalize police work arose from the need to alienate police officers from the class and ethnic groups from which they were recruited. Class and ethnic loyalties made the police ineffective arms of the state. This also gives us an insight into the development of the National Guard as a backup police force (see Balkan et al., 1980: 87).

Traditionally, then, the police institution has been considered a rational response to patterns of disorder either endemic to immigrant groups or characteristic of them as they confront the conditons of a heterogeneous, industrializing, urban society (Lane, 1971; Warner, 1971; Richardson, 1970). This point of view, however, neglects the expanding role of the police (and of legal institutions generally), during America's most intense urban and industrial development. It neglects data that show the police institution expanded most when serious crime rates steadily were falling. Harring (1976) for instance, points out that the expansion of the Buffalo police department between 1872 and 1900 had little to do with an increase in crime. Rather, it was directly related to labor organization and demands for economic concessions from industrial interests.

The account of the origin of the police as a reaction to the criminality and general antisocial behavior of immigrant groups ignores another fact. This is the dynamics of emergent capitalism in leveling and atomizing individuals and that also served to undermine traditional social institutions and modes of informal control within these groups (i.e., the family, the church, the community). It resulted in a docile, homogenous, group of consumers[1] (Spitzer, 1981: 328). The interethnic competition often fostered by the business community to prevent labor organization also had the effect of rewarding "Americanized" behaviors at the expense of traditional authority structures. Proud ethnics, seeking to avoid further stigma, often volunteered to police themselves in ways that were not traditional for them, thus further undermining former authority structures.

In 1908, for example, as the result of an article published by Theodore Bingham, New York City's police commissioner, attention was drawn to a Jewish crime wave then plaguing the city. Bingham claimed that Jews accounted for as much as half the crime in New York. The article unleashed a storm of protest from an outraged old immigrant and native Jewish community, forcing Bingham from office. Jewish leaders, however, were forced to admit that crime was a serious problem, especially among their youth. A Jewish newspaper began gathering statistical data to refute the Bingham charge that half the criminals in New York were Jewish. The headlines of the paper triumphantly proclaimed that although Jews constituted more than a third of Manhattan's population, they accounted for "only" 27 or 28 percent of night court prostitutes. The growing concern within the Jewish community caused middle- and upper-class Jews living uptown to found the Kehillah movement to assist the city police in controlling Jews in the Lower East Side by providing them with information about illicit activities (Steinberg, 1981: 112-114).

The history of the founding of the police, then, clearly indicates that although police actions and organization traditionally has been, and continues to be, conceived of as shaped by the behaviors of the lower class (and today sometimes underclass) and, often, minority groups it controls, the tendency to view that relationship in this fashion is a direct result of the intellectual history peculiar to American ethnics. Policymakers and members of the general population alike from the nineteenth century forward conceived of the members of such groups as innately inferior, crime-ridden, and dangerous. What we have attempted to demonstrate is the other side of what is truly a dialectical relationship. It is just as reasonable (perhaps more reasonable) to conceive of the crime of ethnics from the nineteenth century forward as (at the very least) shaped by the agents of control within the context of a capitalist system.

CRIMINAL JUSTICE AND IMMIGRATION POLICY: THE TWENTIETH CENTURY

We argue that the ideological threads that shaped attitudes and policy in the nineteenth century continue to shape them in the twentieth century. Based on the Dillingham Commission's work, the Johnson Act of 1921 mandated a restrictive immigration policy based on quotas. The National Origins Quota Act of 1924 was based primarily upon testimony of Dr. Harry Laughlin, a eugenics expert of the Carnegie Institution, before the House Committee on Immigration and Naturalization two years earlier. Using highly questionable methodology, Laughlin examined the distribution of ethnics in 445 state and federal institutions. He produced a ranking of ethnic groups by their propensity to insanity, epilepsy, feeble-mindedness, dependency, tuberculosis, and *crime* (Handlin, 1948: 77-78, 104-109).

From about 1923 to 1948 a set of competing factors related to views on the criminality of ethnics and immigration policy emerged. The importance of *environmental factors* began to be recognized. The National Commission on Law Observance and Enforcement's "Report on Crime and the Foreign Born" (the Wickersham Commission, 1931) concluded that the foreign-born were less criminalistic than the native population. Handlin (1948: 141) observes that newer developments in genetics, anthropology, sociology, and other social sciences discredited earlier racist explanations and that "by 1940 it was difficult to find a serious, reputable American exponent of the racist views so widely held."

Curiously, during part of this period of enlightenment, between World War I and 1930, crime was on the upswing (except for homicide, which showed a steady decline from 1920 to 1928; Gillin, 1935: 28). The Volstead Act, which went into effect in 1920 and prohibited the manufacture and sale of alcoholic beverages, certainly explains much of this increase. It is with this act that we witness the birth of organized crime. It is perhaps testimony to the power of organized crime, the collusion of the urban political machine, and desire of established elite to have their drink that the outrage of some over the booming business of organized crime among ethics was not translated into renewed attempts to restrict immigration even further. With regard to these particularly important years in the history of organized crime, Bequai (1979: 34) says that

> the pre-Prohibition era can be compared to the Dark Ages. It was a period of numerous warring gangs. . . .The underworld was highly fragmented, torn by ethnic animosities and rivalries over territories. Prohibition required the underworld to supply a national market and deliver large supplies of liquor to hundreds of thousands of speakeasies.

The official crime rate dipped following the collapse of the economy in 1929 and the repeal of the Volstead Act in 1933. This was most likely due to the sense of community engendered by the Great Depression; the passage of the Wagner Act, which gave trade unions a legal right to bargain collectively and curtailed labor violence; and the New Deal social reforms of Roosevelt, which tended to make racial and ethnic groups feel part of the American fabric (Silberman, 1978: 30). It was only after World War II that the crime rate began the steady surge that has continued to the present.

The decline in the official crime rate prior to World War II may be explained by a combination of several factors. One is the increased interethnic accommodation and specialization within organized crime. A second is the more sophisticated organizational forms and cooptation strategies on the part of ethnics now well-entrenched in *crime as work*. Despite a series of investigations to uncover such criminal activities as well as the complicity of agents of social control (among these the Lexow investigations of the 1890s, the Seabury investigations of the 1930s, and the Kefauver investigations of the 1950s), it is clear that such work, to the degree that the worker was successful, was coming more and more to be seen as at least marginally legitimate. Bell (1960: 117) notes that

> the desires satisfied in extra-legal fashion were more than a hunger for the "forbidden fruits" of conventional morality. They also involved the complex and ever shifting structure of group, class, and ethnic stratification, . . . such normal goals as independence through a business of one's own, and such "moral" aspirations as the desire for social advancement and social prestige. Indeed, it is not too much to say that the whole

question of organized crime in America cannot be understood unless one appreciates (1) the distinctive role of organized gambling as a function of a mass-consumption economy; (2) the specific role of various immigrant groups as they, one after another, become involved in marginal business and crime; and (3) the relation of crime to the changing character of the urban political machine.

Immigration policy past World War II officially remained rigid. The restrictions formulated during the 1920s were the rule. The 1952 revision and codification of immigration law (McCarran-Walter Act) followed old-line policy rather closely and, in fact, accentuated the injustices of the policy. It was certainly affected by the Kefauver investigations. The actual reception, however, that was accorded new arrivals after World War II was more gracious than that of early twentieth century immigrants.

This more recent history marks a change in attitude, then, a decline in the automatic assumption of negative characteristics of immigrants and a certain, albeit subterranean, acceptance of ethnic involvement (and criminal justice collusion) in organized crime as simply another example of American business initiative and acumen. Ideology, economy, legal and illegal work, and social control efforts interact in this period differently than in the nineteenth century, but the importance of the interactions nevertheless is undeniable. In short, at least some of the "new" immigrants have "a leg up."

ETHNICITY, CRIME, AND CRIMINAL JUSTICE: A TWENTIETH-CENTURY SOCIAL SCIENTIFIC BLIND SPOT

Although the discussion thus far has focused on the relationship between an emergent criminal justice system and the criminality of ethnics in the nineteenth century, we would argue that the sort of focus suggested here could fruitfully be applied to contemporary ethnic and racial groups and that it is rather curious that it hasn't. That is, analysis of the criminality of the newest immigrants from Vietnam and Hong Kong on the West Coast as well as Cubans and Haitians on the East Coast, and of newly urban Native American peoples in the cities of the Midwest as well as undocumented Mexicans in the Southwest would benefit from a historical reading that takes fully into account the mutual shaping of organizations of social control by ethnically specific crime patterns and the social, economic, and political conditions of the lives of ethnics and vice versa. The major additional emphasis is that today it is often not sufficient to examine these interrelationships against the backdrop of the conditions and demands of the U.S. economy alone. In more and more cases, examinations of the criminality of ethnics as posed above require one to consider the relevant world economy. For example, when looking at patterns of involvement in drug trafficking among new immigrants from Southeast Asia and of intraethnic extortion, one must consider the worldwide political economy of heroin production, sale, and use. Special attention must be paid to the effect of American law and enforcement strategies on these patterns as well as U.S. involvement in drug-related activities during and since the time of our infamous war in Southeast Asia (see, for example, Chambliss, 1978). For the most part, however, such analytical schemes are absent from contemporary studies of the criminality of ethnics. We suggest that this is due to the rise of social science as a relatively conscious shaper of public policy and to the militant stance of representatives of some new ethnics, especially during the latter part of the century.

Early sociological portrayals of crime in general studies of immigrant and racial groups done at the University of Chicago set the stage for the development of the sort of research proposed here. In the 1920s Park and Miller (1969) offered a relatively evenhanded treatment of the Black Hand, for example, and Thrasher (1963: 145) attributed the rise of the tong among the Chinese in America directly to the need for Chinese labor. Perhaps more important, these early researchers tended to be well aware of the impact of poverty in shaping the crime of new ethnics as well as racial minorities. They tended (quite properly, we would argue) to conceive of much of the criminality of these people as *work*. Their research foreshadowed a distinct theme among social scientists that continued through the first half of the century, but has virtually disappeared from general studies of minorities published more recently. Even economists (whether of the Right, such as Thomas Sowell, or the Left, such as Earl Ofari) do not mention crime as business in the ghetto. Bettylou Valentine's (1978) monograph on hustling as the work of underclass Blacks is unique in this regard.

Many studies have emphasized the oppressiveness of the criminal justice system with regard to specific minority groups and, sadly, many have continued to stereotype their criminal involvement. Very little exploration of the amount of involvement of ethnics in illegal work or of the various patterns of that involvement has been done. There are several reasons for this state of affairs. First, relatively assimilated, upwardly mobile ethnics have often discouraged such research, either denying that such patterns exist or arguing that such studies support discriminatory action against ethnics by fostering images of the crime-prone newcomer. The traditionally liberal social scientist responding to such arguments or anticipating them has either avoided in-depth studies of such groups or generalized with insufficient data. Thus, ironically, already existent stereotypes are often perpetuated or new ones generated. These new stereotypes often reflect more positively on the groups themselves, but are in their own way just as unfair and damaging, not to mention intellectually unenlightening and dangerous with regard to public policy formation. Most important, the feeling that to study crime among ethnics is to stigmatize them has often been held most strongly by ethnic researchers themselves. Thus even those most likely to have the fewest barriers to carrying out such studies have often been reluctant to do so.

Last, it is interesting to note that this neglect occurs during the very period when several brands of criminology of the Left (radical criminology, Marxist criminology, the New Criminology, and critical criminology) were stimulating interest in examining criminal patterns within the broad context of historical developments in capitalist economic forms and legal superstructures. The reason for ignoring the criminality of ethnics and the relationship between ethnicity and the criminal justice system generally is twofold. First, especially in its most naive forms, conflict criminology has also been guilty of ignoring the very real criminality of proletarian groups because its advocates fear that such studies would legitimize increased coercive measures against those least empowered to resist. There is also the very clear message in some of the earliest work of these researchers that somehow the crime patterns of the proletariate are less troublesome, more reasonable, more politically correct than the crimes of capital. Some radical scholars simply wish to refocus criminology on the much-neglected crime of capitalist and corporate structures with workers painted simply as victims. Second, class-based conflict criminology is, by its very nature, just that—class based. It tends to ignore (or regard as dangerously divisive) analyses that focus upon individual ethnic groups. It is ironically perhaps a sad commentary on the stigmatized and truly powerless status of urban underclass Blacks that they have been more the focus of the sorts of studies advocated here than have contemporary ethnic newcomers (see, for example, Valentine, 1978; Ianni, 1974).

STUDYING THE NEWEST IMMIGRANTS

There are obviously numerous historical studies along the lines suggested above that might be done. There are also studies to be done of newly arriving members of groups that have been in the United States for substantial periods of time. Such studies surely require a certain historical vision as well. If current research trends continue, however, a most important opportunity will be lost: the chance to study the newest new immigrants as they arrive (and/or are selectively returned to their countries of origin), attempt to carve out a life in the United States, and have those attempts molded by stereotypes, concrete material conditions, and the exigencies of the economy. These studies need not be historical ones (although history will obviously always be important to such research). The increased understanding of how the variables cited here interact in human terms should not be foregone.

To those who argue that such studies do a disservice to immigrant groups already under much stress, we would only point to the alternatives that already are in operation. In the past fifteen years or so, policy with regard to both crime and immigration has been increasingly cast in cost-benefit terms. Banfield (1974), Wilson (1975), and others have suggested that the United States should not concern itself with issues of etiology, that this needlessly mires the policy analyst in essentially irrelevant issues. What should be of concern, they argue, is what appears to work: what would reduce property crime and how it can be achieved as cost-effectively as possible, for example, incapacitation through mandatory imprisonment. Thus, to the degree that ethnicity is considered in criminal justice policymaking today, it is as an aid to projecting the future burden of such groups on already heavily taxed human services (see, for example, U.S. Department of Justice 1980). The toll in human terms of such policies is ignored as is the fact that such policies often impact on ethnic groups in a way that further increases the burden of the group on other human services by undermining community supports and informal controls.

It is no longer fashionable in most academic circles to speak of the genetic inferiority of particular racial or ethnic groups. To do so one must be as careful as conservatives Miller (1958), Banfield (1974), and others when they speak of the problems of "lower-class" peoples. Yet, there very clearly has been a renewed interest among some criminologists in biological determinism (e.g., Mednick and Volavaka, 1980; Jeffrey, 1978). Public policy has reflected this conservative trend. Thus, in 1971, Arnold Hutschnecker, one of President Nixon's personal medical advisers, proposed a massive program for the chromosomal screening of every 6-year-old in the country. He recommended that "hard-core" 6-year-olds demonstrating evidence of criminal tendencies be sent to "therapeutic" camps to learn to be "good social animals." This particular plan was sent to Eliott Richardson, then secretary of Health, Education and Welfare, who turned it down "because it was not feasible to implement on a national scale" at that time (Hunt, 1973; Moran, 1978: 347).

Consequently, those who are reluctant to study the criminality of the newest new immigrants should be aware that they do those peoples no favors. Conservative researchers with other agendas will fill the void with research that, while ignoring the uniqueness of the ethnic groups themselves, will appear relevant to the formation of public policy that affects them. Policymakers will forge ahead with few data. The findings that could challenge such trends simply will not exist.

The reality that needs to be investigated began in the 1950s and it continues. By the 1960s it was clear that the 1924 National Origins Act was at the very least poorly targeted. First of all, the immigrants from southern and eastern Europe especially were generally thought at this time to be desirable newcomers because of their status as political refugees. Continued discrimination against them made little sense when they obviously had much to contribute, particularly during a period of Cold War build-up. Second, the number of immigrants from Latin America had swelled. These were not thought to be desirable newcomers. The problem was that the 1924 legislation did not limit their entry. The Immigration Act of 1965 sought to rectify this situation. In other words, it sought to ensure that those considered less desirable were restricted and to mandate a preference for professionals and the technically trained.

This legislation encouraged a significant number of highly skilled Asians to immigrate at the same time that it "criminalized" an enormous group of undocumented Latin (particularly Mexican) newcomers. But the push of political upheaval and growing industrialization in the traditionally agricultural native lands of these latter prompts them to continue to come during this period as does the pull of the desires of American manufacturers and agribusiness for cheap labor (enter the Simpson-Mazzoli bill).

These groups then should certainly receive the attention of researchers. But with this prior immigration as background, other new newcomers also deserve attention. For example, despite the historic existence of tongs in America, low crime and delinquency rates among descendants of Chinese and Japanese immigrants to the United States have repeatedly been noted (e.g., Sollenberger, 1968). In the 1960s, however, Kitano found some interesting signs of "contamination" when he compared Japanese American delinquents and their parents with Japanese American nondelinquents.

In the 1970s there were reports that new, young, and unskilled male immigrants from south China who had settled in Chinese ghettos in several American cities were engaged in gang-related warfare (Harvey, 1970; Petersen, 1972). Most recently, President Reagan's Commission on Organized Crime heard three days of testimony related to Chinese and Japanese criminal organizations in the United States that purportedly have ties to traditional organized crime here and in Hong Kong and Taiwan. It was reported that Chinese secret organizations, called "triads," control the importing of heroin from Southeast Asia's Golden Triangle and are responsible for 20 percent of the heroin now smuggled into the United States. According to the witnesses, besides trafficking in heroin and cocaine, the Chinese are involved in extortion, gambling, robberies, and prostitution. They are also said to have bought large numbers of guns from "the Mafia" and to employ large groups of refugees from Southeast Asia, especially Vietnam, because of their experience in handling guns (Raab, 1984: 1, 16).

Chang Pao-Min (1981) reports that crime has become an increasingly serious problem among the Chinese. Between 1960 to 1975 the increase in the total volume of crime was 98.5 percent for the entire country, 2.2 percent for the Japanese, and 157.2 percent for the Chinese. Although the Chinese population is one-fourth smaller than the Japanese, the total arrest rate in 1975 was higher than for the Japanese and in violent crimes actually surpassed the Japanese in absolute numbers. Before 1966, the number of Chinese juvenile delinquents in New York's Chinatown never exceeded 10 per year. In 1966, it jumped to more than 200 (Pao-Min, 1981: 363-367).

Both the newest Latin immigrants and the newest Asian immigrants have predecessors in prior waves of immigration. This necessitates a certain historical dimension to research on their criminal activity. The latest wave of Cuban immigrants who came courtesy of Fidel Castro, despite the fact that they are quite different from the Cubans who preceded them, also fall into this category.

Recent attention to what appears to be organized criminal activity on the part of Nigerian immigrants focuses attention on a distinct category of new newcomer, those who, as a group, have no real predecessors. A *New York Times* article documents efforts of federal and local law enforcement authorities to halt rings of Nigerians across the country who apparently specialize in credit card fraud although they are also reported to be involved in trafficking in drugs and guns. The profits from such activities are then purportedly used to buy American consumer goods, which are shipped to Nigeria and sold in the extensive black market there (Schmidt, 1983: 1). Other groups that fall into this latter category include several, such as Haitians, who come from the Caribbean seeking political asylum.

One last category of "ethnics" surely deserves attention. It includes those unique peoples who have been the victim of American colonialism and imperialism: These include some Hispanics and Native American peoples, including Eskimos and those residing in American territories and possessions. We know next to nothing about these peoples with reference to the research questions posed in this chapter.

The research agenda sketched here is a formidable one. Even the initial spadework has really yet to be done. However, both the intellectual and the political significance of the task demand our immediate attention.

NOTE

1. The production of commodities as well as the production of a consumer class also plays a role in the crime patterns of immigrant groups. Eitzen and Timmer (1984: 373) argue, for example, that at the time in question "alcoholic drink [especially by the Irish] was generally believed to be a primary cause of crime and public riot. Ironically, public drunkenness was unknown until the emerging market economy encouraged the commercial sale of alcohol. When this occurred, drinking was torn from communal and family-oriented mealtime use. Alcohol became a commodity for private consumption."

8

Ethnicity and Social Welfare in American Cities: A Historical View

ROBERT S. MAGILL

☐ PEOPLE HAVE HELPED OTHERS in need since before recorded history. Social welfare services have been provided by family networks, religious organizations, voluntary citizens' efforts, and governments. In a modern, technological, urban society, social welfare has become a necessary institution. Even with recent cutbacks, social welfare commands large outlays of time and money at the federal, state, and county levels of governments and by the private sector.

Since social welfare occupies such an important role in American life, it is surprising that so little formal attention has been given to social welfare. Nor until recently has there been much concern for or scholarly interest in urban immigrants. While more and more Hispanics, Native Americans, Asians, and South and Central Americans are immigrating to our central cities, it has been only recently that academics and policymakers have taken notice and begun to describe this extremely important social, political, and demographic change. This chapter presents what we now know about ethnicity and social welfare.[1]

SOCIAL WELFARE: A DEFINITION

Social welfare can be defined as

the organized system of social services and institutions, designed to aid individuals and groups to attain satisfying standards of life and health, and personal and social relationships which permit them to develop their full capacities and to promote their well-being in harmony with the needs of their families and the community [Friedlander, 1961: 4].

Social welfare is a necessary institution in a modern society. In fact, there is no known industrialized society, regardless of its political or economic system, which does not have some form of organized social welfare (Rimlinger, 1971).

Most basic social welfare services are provided by government. The efforts of private agencies supplement government efforts. In 1978, for example, federal, state, and local governments spent $394.46 billion on social welfare, while private agencies spent $154.4 billion (McMilan and Bixby, 1980).

Social welfare services usually are characterized by the unilateral transfer, by the gift or the grant. In contrast to transactions in the private marketplace, consumers of social welfare services usually do not pay the full cost of the services provided (Boulding, 1967: 7). In other words, most social welfare programs are subsidized by the government or by private charity. This subsidizing is important because it means that if a social welfare program is to survive and prosper, it must satisfy the requirements of the granting body, which may or may not support policies that are in the best interests of clients. In contrast to the marketplace, demand for social welfare services does not exert a direct control on the quantity or quality of social welfare services available. In addition, since the client does not pay full cost, there is a potentially unequal relationship between the provider and the consumer of social welfare services. If the provider-consumer relationship is abused, clients can internalize their dependent lower status and become less successful in solving their problems (Kerson, 1978).

Historically, social welfare provisions in the United States have been organized around specific social problems and the needs of particular populations that are at risk. Following this approach, it is clear that more attention must be given now to the needs of the new urban immigrants.

During colonial times, social welfare services were provided by local communities. Help was generous and was seen as part of a religious duty (Trattner, 1979). After the Revolutionary War, when social problems became more serious, more formal social welfare structures were established. During the 1800s, most large American communities built almshouses where all of those who had to be institutionalized, such as the old, the retarded, the emotionally ill, and unprotected children, were housed. There was also a system of outdoor relief for those who could live in the community, but needed supplemental income, work, or counseling.

Social welfare needs soon outstripped the abilities and resources of local communities. In addition, it was recognized that more specialized services were needed for specific groups such as the emotionally ill, the physically handicapped, and those requiring economic assistance. During the middle and the late 1800s, some states took responsibility for providing some social welfare services.

During the Great Depression, it became clear that the federal government had a basic responsibility for the old, the disabled, and poor children. The passing of the Social Security Act by Congress in 1935 was the first time that the federal government assumed major responsibility for the social welfare part of the constitutional directive "to promote the general welfare." This precedent formed the basis for the modern welfare state, although most federal social welfare funds are channeled through the states, the counties, the cities, and private agencies. Social security, medicare, federal penitentiaries, aid to Native Americans, and programs for war veterans are the only major social welfare policies that are operated wholly by the federal government. Which level of government should provide which social welfare programs always has been a major issue in American social welfare policy. Since the 1930s, the federal government has assumed an ever-increasing role in the funding and control of social welfare policy. During the past ten years, this trend has been reversed. The New Federalism of presidents Nixon and Reagan has decreased federal social welfare expenditures and responsibilities in the area of social welfare policy significantly, while encouraging states and counties and the private sector to accept larger responsibilities. Most social welfare experts feel that the consequence of this trend has been to reduce significantly social welfare provisions, especially for the poor and minority groups. The result has been a dramatic increase in the percentage of the population that is poor, ill housed, ill clothed, and ill fed, especially among minority groups (Magill, 1979). The consequences of these cutbacks are especially serious for the new urban immigrants who desperately need help surviving in urban America.

In most Western countries, social welfare policies are primarily the responsibility of the government. However, in the United States, there is a long history of nongovernmental activity. When he studied America in 1783, Frenchman Alexis de Tocqueville (1956) observed that Americans tended to form voluntary associations and solve problems through the private sector.

This pattern of public and private involvement has been very evident in social welfare. Even after the federal government became involved in providing social welfare services, the private sector has remained important. While clear divisions of responsibility do not exist, in general, governmental bodies at the federal, state, and local levels tend to provide the basic provisions in certain areas. These areas include income maintenance, food, housing, medical care; some mental health services for the emotionally ill, disabled war veterans, and the addicted; and various programs for the aged. Governmental programs tend to serve low-income and working-class persons, including many of the new urban immigrants.

On the other hand, private social services tend to provide counseling and other services primarily to lower-middle-, middle-, and upper-class persons. Many observers see a trend toward a social welfare system based on class. Government provides the less generous, stigmatized services to lower-class individuals and minority-group persons, including the new urban immigrants. Many private sector agencies tend to serve primarily higher status clients (Kammerman and Kahn, 1976).

VALUE ORIENTATIONS IN SOCIAL WELFARE POLICY: THE COLLECTIVISTS AND THE INDIVIDUALISTS

As an institution, social welfare is a part of the society and reflects the major values of American society. Social welfare is an area in which people have especially well-established values. This phenomenon occurs in part because social science research has not advanced to a stage at which, with a high probability, it can predict the consequences of various efforts to solve social problems.

A large number of people generally have opinions about most social problems such as poverty, juvenile delinquency, and the care of the elderly. This assumption of adequate knowledge and an informed capacity to make policy is not as true of other areas of public policy, such as the development of specific defense strategies or the setting of priorities and developing delivery systems for medical services. Values, therefore, tend to be more important in social policy than in other areas of public policy.

Some of the most powerful American values seem to be in conflict with basic elements of social policy. Individualism, competition, social mobility, and self-sufficiency conflict with a basic approach of social policy—helping those who are in need. Powerful opinion leaders, such as the economist Milton Friedman (1972) and Ronald Reagan identify as their primary value the freedom of the individual to choose. Friedman and other individualistically oriented policymakers oppose almost all infringements on the value of individual freedom. They particularly oppose efforts toward social justice, which often require that the power of government be used to redistribute resources, create opportunities, or raise taxes. Their opposition is strong because the government's exercise of such power infringes on some individuals' social, political, and economic freedom.

Individualistic orientations always have been important in American society. For instance, Social Darwinism was once a dominant value orientation (for example, see Mirringoff, 1980). This approach is reminiscent of the early English Poor Law prescription of *less eligibility*. Less eligibility required that welfare assistance payments should not be higher than the wages of the lowest paid public employee. Poor Law supporters felt higher payments would result in a disincentive to work. Later, the Reverend Thomas Malthus and others argued that supporting the poor would lead to their geometric increase and a consequent weakening of society. Progress was based on natural evolutionary laws, which, like the market, should not be tampered with. As Charles Darwin wrote, "The natural forces of elimination (of the poor and dependent) served to preserve the best or most 'fit' elements of society. Social welfare institutions, in contrast, function to enfeeble the society through their artificial preservation of 'unfit' species" (quoted in Mirringoff, 1980: 304).

Historically and currently, it is possible to identify two major value orientations regarding the role of social policy in society. They are the collectivist orientation and the individualistic orientation. These two orientations are really opposite ends of a continuum. Many persons combine in their thinking some elements of both the collectivist and individualistic orientation.

In general, individualists believe that economic, social, and emotional problems are based on individual deficiencies that can be corrected in a free society by the individual. Individualists tend to, in William Ryan's (1971) famous phrase, "blame the victim."

Collectivists feel that the causes of an individual's problems are related to a complex interaction among many factors, including heredity, the family. the community, the school, and the values and institutions in the society. Collectivists feel that values and organizations in society have an important role in creating social problems, which then became the target of social policies.

To solve social problems, individualists place major responsibility on the individual suffering the problem. For example, individualists feel that the poor should be able to "pull themselves up by their own bootstraps."

The degree and nature of governmental participation in social policy is one of the major issues that distinguishes individualists from collectivists. Traditionally, collectivists have looked to government in general, and the federal government in particular, as an appropriate vehicle for social policy. Individualists have resisted all but minimal participation by government. They fear that too much reliance on government to solve individual problems can weaken freedom, the market form of economy, and individual self-reliance.

When government is needed, individualists generally have favored the delivery of public services by the lowest level of government. Preference is given to municipal government over state government and states' rights before federal involvement (Shick, 1969).

Individualist values tend to dominate American society, although there have been two periods, the 1930s and the 1960s, when the collectivist orientation was dominant. Not much is known about the value systems of the new immigrant groups. Many are from urban environments, but they also include groups like the Hmong which have been airlifted directly from very simple societies in which strong values were placed on self-sufficiency. There may be some value clash between the individualistic orientations of many of the newer urban immigrants and their needs for provisions based on a collectivist philosophy. As these groups become more acculturated within American society, and it becomes clear that survival is more complex in the city than in rural areas and mountains, these individualistic values may change.

THE FUNCTION OF
SOCIAL WELFARE IN AMERICAN SOCIETY

Social welfare is an essential institution in American society and reflects the major value orientations of the society. Its scope and specific areas of concern change over time.

With industrialization, urbanization, and the division of labor, societies have become larger and more complex. Change has become more rapid. The old organic whole of the preindustrial society has been replaced by less clear and logical relationships among the parts of societies. Different institutions, such as the family, government, religion, and education, change, often independently of each other, and social needs are left unmet. A major function of social welfare in an industrial society is to help those whose problems have been caused by the complexity and institutional conflict characteristic of industrial society.

In addition, modern societies have conflicts between generally prescribed values and institutions. For example, it is clear that the general values and goals of the society, such as economic success, are not available to all members of the society. Success may be the result of hard work, chance, or a special advantage that some members of the society enjoy and others do not. As Robert Merton (1968) writes, "When a system of cultural values extols, virtually above all else, certain common success goals for the population at large, while the social structure rigorously restricts or completely closes access to approved modes of reaching these goals for a considerable part of the population, deviant behavior ensues on a large scale."

Social welfare has been given the responsibility for rationalizing this disjunction between generally accepted goals and differentially distributed means. Social welfare's function in this context is to develop ameliorative policies and to deal with the deviant behavior that results when policies are not effective.

A second function of social welfare in a capitalist society is to provide for those who are unable to survive in the private economy. With the breakup of feudalism, a rigid class structure was replaced by an economic system that emphasized private gain. This resulted in significant social and economic inequalities. This early characteristic of the market system—to create classes based on income and wealth—has continued to the present.

Further, a pure market system makes no provision for those such as children, the aged, the handicapped, and the emotionally ill, who, for a number of different reasons, are not taken care of by parents or relatives and are unable to work. For those who are unable to work, or who are paid less than a living wage, the market system, by itself, is unfeeling and brutal and needs to be supplemented with social welfare policies. A market economy can create significant social and economic inequities, and some, such as Richard Titmuss (1965), the famous British social theorist, feel social welfare should be used to keep inequalities from becoming too great.

Social policies can have an important and direct impact on the size of the labor force. During Elizabethan Poor Law times, social policies were enacted to force individuals to work. In general, then and now, social policies tend to be punitive toward workers when there is a large supply of low-skilled workers who are not attracted by low-paying and low-status jobs, which usually have undesirable working conditions. Social policies tend to be supportive of workers who are skilled and who are difficult to replace (Magill, 1984).

At times, government pursues conscious policies designed to cause unemployment in order to increase the size of the labor supply and thereby lower inflation. In past recessions, since the Great Depression, government has maintained social welfare policies, such as unemployment insurance, in order to preserve social stability. During the most recent recession, the Reagan administration reduced social welfare benefits. This reduction made the risk of unemployment more serious, put downward pressure on wages, and forced workers to accept less desirable working conditions. Many feel that the creation of higher unemployment and the simultaneous reduction of social welfare programs were orchestrated purposefully to increase the size, and therefore reduce the demands, of the blue-collar labor force. As Piven and Cloward (1981: 466-467) write, "Conservatives believe that, in an industrial society, aid to the needy reduces business profits by enhancing the bargaining power of the labor forces. . . . Slashing social programs . . . reinstate(s) the terrors of being without a job."

Social welfare is concerned with both social change and social control. At times, social welfare has supported major efforts toward social change. This support of change was true of the progressive efforts of the early 1900s. Social welfare made a major contribution to the New Deal of Franklin Roosevelt and to the Great Society of Lyndon Johnson. The social welfare policies of Franklin Roosevelt significantly reduced inequality in society. They improved the lives of the elderly, women, children, and the unemployed. Similarly, during the Great Society period of John F. Kennedy and Lyndon Johnson, there were significant advances for those in need.

Piven and Cloward (1982) argue that social welfare has expanded to such an extent that it has changed permanently the relationship between labor and capital in the United States and is therefore fulfilling more of a social change function.

President Reagan proposed to cut $140 billion from social programs between 1982 and 1984. More than half of this cut came from income maintenance programs that provided the poor with money, food, housing, and health care. According to Piven and Cloward (1982: 31), "The income maintenance programs are coming under assault because they limit profits by enlarging the bargaining power of workers with employers. . . . the income-maintenance programs have weakened capital's ability to depress wages by means of economic insecurity, especially by means of manipulating the relative numbers of people searching for work. In effect, these programs have altered the terms of the struggle between business and labor."

For Piven and Cloward, increasing social welfare benefits, such as unemployment and disability insurance and other income maintenance programs, have reduced somewhat the size of the low-skilled labor market and made the terrors of unemployment less serious. Because of structural changes in the economy and the popularity of Keynesian thought, the role of government in the economy has become more obvious. Workers now hold politicians accountable for their individual economic situations, according to Piven and Cloward (1982: 125): "A century and a half after the achievement of formal democratic rights, the state has finally become the main area of class conflict." In contrast to many other observers, Piven and Cloward assert that social welfare policy has been effective in improving the condition of the poor and of workers, and they anticipate that it will become more important in this area in the future.

On a general level, it can be argued that the ultimate aim and the consequence of social welfare policy are social stability and, therefore, social control. Many of the programs of the New Deal provided the alienated with a stake in society and may have forestalled more basic social and economic change. Similarly, the Great Society programs were able to provide a legitimate avenue for the release of tension. By providing new opportunity structures, they revived the Horatio Alger myth, which had been exposed as a fraud for minority Americans. By hiring and thus co-opting the agitators and redirecting their efforts toward acceptable political conflict, these programs successfully maintained the basic parts of the existing institutional framework.

Social change is dominant in social welfare policies and programs when there are basic and powerful domestic threats to the society such as the Depression and the civil rights revolution. Social change is supported also when, because of demographic and other changes, particular groups in need are able to gather support for their problems. For example, over the past decade, the elderly have been particularly effective in attracting support for their problems.

Social control has been dominant in social welfare policies and programs when the society is relatively stable and the existing political processes can handle most conflicts. Social control predominates also when specific groups in the population are unable to gather support for their needs.

From another perspective, social change and social control in social welfare are part of a recurring pattern. As social problems develop and become significant, social welfare organizations make efforts toward bringing about social change. If successful, these efforts channel the social change within acceptable limits and in the process may provide important assistance to those in need. This assistance relieves the pressure for more basic change. In time, the successful social welfare policies are curtailed. A social control approach prevails. When the political and economic systems cannot accommodate the needs created by these policies, pressure eventually builds for new social welfare policies.

For the past ten years social welfare, on a general level, has functioned as a social control mechanism. It has acted both as a Band-Aid and a valve to relieve pressure. During this period there have been serious cutbacks in social welfare services. The prevailing ideology, individualism, has worked against social reform. While some specific individuals have been helped by some social welfare programs, progress has not been made for groups. During this period there has been a dramatic increase of those in need, including the poor, children, single-parent families, abused women, displaced homemakers, and the newer ethnics who have migrated to the cities. Many of the gains of the 1960s have been lost. The result is that a larger proportion of the population is now in need than was in need ten years ago. This resurgence of need provides the opportunity for a reemergence of social welfare as an important tool in the quest for social justice for the population in general and for the newer urban ethnics in particular.

SOCIAL WELFARE AND EARLIER IMMIGRANTS

As the United States became industrialized, in the absence of federal social welfare policies, formal, private sector social welfare organizations became more prominent. With the growth of cities and an increase in the immigration of eastern Europeans, American society was facing new problems. The Charity Organization Societies and the settlement houses started in the late 1800s to serve the poor and needy in growing central cities. Many of those who received help were eastern European immigrants.

The Charity Organization Societies were the predecessors of our modern family counseling agencies. They worked with individuals and families to provide financial and other assistance to those in need. The Charity Organization Societies were supported totally by private donations. They tended to follow an individualistic philosophy, blaming clients for their problems and helping them to adjust to existing social and economic conditions. Since there were few state- and local government-supported public welfare programs, the Charity Organization Societies fulfilled an important role in helping some of those in need to subsist in an urban, industrial community (Woodroofe, 1971: 25-55).

In contrast, the settlement houses worked with groups of people and had a more collectivist philosophy. While providing to immigrants direct services such as instruction in the English language, health and nutrition education, child care, and job training, settlement house leaders worked to change social conditions on local, state, and national levels.

The new immigrants were a threat to many Americans, and they were treated with prejudice. Organized labor feared that immigrants would work at reduced wages and in unsafe conditions. Others feared that the new immigrants from eastern Europe were racially inferior and that they would bring with them crime, disease, poverty, alcoholism, promiscuity, and slums (Bremmer, 1960).

Middle-class reformers started to speak out in defense of the new immigrants. A Harvard-educated philosopher, Horace Kallen, supported cultural diversity as essential to democracy and defended the right of ethnic groups to preserve their language, religion, communal institutions, and culture. This view was shared by Jane Addams and other settlement house workers who "learned to appreciate the cultural heritages of immigrants and began to defend their uniqueness" (Colburn and Pozzeta, 1979: 6-7).

According to Allen Davis (1967), the essence of the settlement house approach to the immigrants was to understand the customs and traditions of each group and then to seek as much opportunity for them as possible. For their part, the immigrants were confused and often ambivalent about the help from "outsiders." Like the newer urban immigrants of today, they often had an individualistic philosophy that worked against their achievement of common group needs.

An example of efforts by the settlement houses was the Immigrant Protective League organized by Hull House in Chicago in 1908. The league sought to ease the adjustment of newcomers to strange cities by helping them find housing and employment.

The settlement house movement tended to be liberal rather than radical. It worked within the basic economic and political assumptions of American society, although it did work with others for new policies to protect immigrants, workers, minorities, children, and women. While supporting the uniqueness of eastern European ethnics, settlement house workers also pushed for the assimilation of immigrants into the social, economic, and political life of American cities.

A parallel immigration took place between 1950 and 1960. During this time 1,400,000 Blacks left the South and migrated to cities in the North, Midwest, and West. This movement placed serious pressure on the economies and public welfare systems of these cities. For example, in 1953 unemployment was 20 percent higher among Blacks than among Whites. Ten years later, the average differentials rose to 112 percent (Piven and Cloward, 1971).

The urban social welfare system had difficulty in adjusting to the new clients. The Charity Organization Societies largely had ceased giving out money to the poor when the federal government and the states took over the public assistance function. These agencies had become family counseling efforts, were often sectarian in outlook, and tended to serve working- and middle-class clients. The old settlements were located in zones of transition, but were having difficulty in changing over their staffing and programming patterns from White ethnics to Black residents. Governmental programs, primarily public welfare, were based on a cost-conscious and individualist punitive ideology and were, at best, merely maintaining individuals and families at subsistence levels. Essential social welfare services in the area of mental and physical health, child welfare, supplementary economic assistance and employment training, and programs for those with special educational needs were inadequate or nonexistent. Adding to these problems was the pervasive race prejudice and the difficulty skin color presented in assimilation efforts.

The seeds for major social disruption had been building for generations. Inadequate social conditions led to the civil rights movement and the urban riots. The reaction of the society was to increase dramatically a wide variety of social welfare efforts. In an extremely short time during the beginning of Lyndon Johnson's presidency, new federal policies were enacted in the areas of civil rights, poverty, mental health, physical health, employment, education, and social services. The Community Action Program succeeded in stimulating citizen partici-pation in legitimate political processes.

While inadequately funded, the programs of the Great Society were successful in ameliorating social problems. The rate of poverty decreased dramatically. Inner-city Blacks had better physical and emotional health. Crime rates decreased and employment improved. Unem-ployment among the Black middle class declined dramatically. When the private sector was unable to respond to the social and economic demands, the government took over and created jobs, opportunities, and social support.

This massive effort clearly helped individuals and inner-city Blacks as a group. It also had the effect of "cooling out" the revolution. Many of the civil rights leaders of the 1960s soon became the social welfare, governmental, and political employees of the 1970s.

While many successes could be claimed by the Great Society programs, as in many "revolutions," the leaders seemed to gain more than the followers. As Steven Erie and Michael K. Brown (1979: 6) write, "From the mid-1960s to the early 1970s, there was a correlative increase in the number of social welfare professionals and managers on state and local payrolls and in 'soft' federal grant programs providing social and health services. . . . The fledgling Black middle class has become much more involved relative to Whites in providing health and social services to the disadvantaged."

The 1960s, like the 1930s before them, are instructive in examining the function of social welfare in the face of serious social instability. On the one hand, extensive social welfare programs do help those in need. Basic conditions of living can be improved. Critics of these efforts charge that one result is that clients become dependent, at great cost to society (Erie and Brown, 1979). The reality is not so simple. Clearly, there are some who are so physically or emotionally ill or so young or so old that they will need long periods of assistance from society through social welfare policies. But for many, social welfare programs provide that extra help that is needed in special situations. In contrast to popular myth, the average stay on Aid to Families with Dependent Children (AFDC) rolls is little more than a year. Almost half of AFDC recipients use the program to get through a crisis and then become independent. The families of all our recent presidents, including Kennedy, Johnson, Ford, Carter, Nixon, and Reagan, have all benefited from social welfare programs. These men and women were not made totally dependent on the public dole.

What is more likely, however, is that the benefits of social policies are distributed differentially throughout the population in need. While all may benefit, some benefit more than others. Social welfare, as a profession, always has been attractive to minorities who were barred from other occupations. For a long time, social work, along with nursing, was seen as "women's work." The early workers for the Charity Organization Societies and the settlement houses were, for the most part, women. During the Depression, Italians and, especially, Jews were able to find employment in social welfare. More recently, Blacks, Chicanos, Native Americans, and Asians have found career possibilities in social welfare.

In summary, then, social welfare, at times of extreme social instability, works for change. It tries to lessen the inequities in society that produce social tension. It tries to rationalize deep conflicts such as those between equal opportunity and discrimination or between the rich and the poor. The goal of social change is social control. By reducing the tension in society, social stability is preserved, although the resulting equilibrium is somewhat different. While various groups benefit in this process, the benefits are distributed differentially. Researchers such as Erie and Brown (1979) feel that often the upward-striving leaders who have been creating the social instability are co-opted by the system and receive the most benefits.

THE NEW ETHNICS TODAY

"To live in one of these foreign communities (within American cities) is actually to live on foreign soil. The thoughts, feelings, and traditions. . . . are often entirely alien to an American. The newspapers, the literature, the ideals, the passions, the things which agitate the community, are unknown to us except in fragments." This observation was written in 1904 by the progressive reformer Robert Hunter. To a large extent, it is applicable today to the Asian, Hispanic, and Native American urban newcomers.

Immigrants are voluntary and involuntary. Some come to America of their own free will to seek better economic, political, or religious conditions. Some are forced here as refugees, expelled expatriates, prisoners of war, or slaves.

Distinctions also can be made among immigrants who are political refugees, those who are admitted to the United States under a quota system, and undocumented aliens. Political refugees often emigrated from countries where their social position was greater than it is in the United States. Their expectations may be greater and their demands and frustrations higher than those of other immigrants.

Immigrants admitted under a quota system are likely to have family and social support already available in the United States. Of the three groups of immigrants, they are most likely to be able to succeed on their own, without extensive supplementary supports.

Those most in need are probably the undocumented aliens, who come to the United States out of desperation. Given their difficult status and vulnerability, their need for a total range of economic and social welfare programs is extreme. Yet they are probably least able to obtain the help they need.

Some immigrants become permanent residents of American cities. Others, however (refugees, for example) typically have no real hope of returning to their homeland. In many cases families have been separated, and the new immigrants can feel isolated and alienated. Others are more likely to return home regularly, whether to Puerto Rico or to an American Indian reservation. Old family and friendship ties are reestablished and the returnees can experience "emotional refueling" (George Nakama, personal communication, July 2, 1984). These groups tend to see their urban American experiences as more transitory, and this perception undoubtedly distinguishes their outlook and needs from those of the more permanent urban immigrants.

Typically, the immigrants to the United States and inmigrants to the city face four facets of poverty. They usually have to start at the lowest economic level and usually have to live in extreme poverty. They are uprooted socially, frequently having lost permanently their previous status. Often, the language of the country is different, and employment opportunities are small. Finally, the immigrant to the city often feels isolated and estranged, in conflict with new ways of living, new ethical standards, and a foreign culture (Meyer, 1984: 99).

However, "American society accepted immigrants but never approved of the ethnic groups." (Olejarazyk, 1971: 284). While the culture has promoted the values of assimilation and acculturation, assimilation and acculturation have not occurred, especially for the newer immigrants. "The final blow to the assimilationist ideology has been delivered by the Black American. He could not disappear and therefore did not fit into the concept of American adjustment and consequently had to be put outside the Constitution" (Kolm, 1971). "It is now plain that the concepts of 'acculturation,' of 'assimilation,' and similar paradigms are inappropriate for groups who entered American society not as volunteer immigrants but through some form of involuntary relationship" (Moore, 1976: 131). Using the earlier distinction between voluntary and involuntary immigration, many of the new ethnics—the Spanish speaking, Asian, and the Native American—can be designated as involuntary immigrants.

Generalizing always should be approached with caution, especially when considering the urban immigrants. In fact, they make up a very disparate group, with separate languages, cultures, and interests. In addition, at least in the availability and utilization of social welfare services, there is a serious lack of empirical evidence. Clearly, further study is needed in this area.

While many of the urban immigrants have come from rural backgrounds, they now are flocking to America's cities. As early as 1970, all these minorities were predominantly city dwellers; 81 percent of Mexican Americans, 95 percent of Puerto Ricans living in the United States, and 50 percent of American Indians lived in central cities by the mid-1970s. When they move to urban areas, these groups tend to cluster in separate ethnic communities, in zones of transition known for their inadequate housing, services, and job opportunities (see Moore, 1981).

Many recent Asian Americans were resettled under the Refugee Act of 1980, which was an amendment to the Immigration and Refugee Assistance Act of 1962. This act established a uniform policy for immigration and provided for domestic resettlement assistance, which is now the same for all groups of refugees. The legislation created an office of the U.S. Coordinator for Refugee Affairs at the rank of ambassador-at-large, located in the U.S. Department of State. An office of Refugee Resettlement in the U.S. Department of Health and Human Services was created to administer programs of reception and placement of refugees, including resettlement services (Sullivan, 1978: 172).

Initially, many Asian American refugees were dispersed throughout the country and settled in small American communities. Since they were often isolated from others of the same culture, the adjustment was extremely difficult (Wong, 1983). At the beginning of the major immigration, Puerto Ricans also were settled in smaller communities and employed in low-paying jobs (Weil, 1983). Both groups eventually moved to larger urban communities and settled in neighborhoods with others of their cultural group.

Some of the new immigrants, including Asians and Hispanics, are professional persons, including doctors, lawyers, engineers, artists, and teachers, who have difficulty in adjusting and in finding acceptable employment in the United States. However, many of the new urban immigrants are unskilled and poorly educated.

Other chapters in this volume address the specifics of poverty and the labor market status and prospects of the new immigrants. They show an enormously complex picture, one in which economic need tends to predominate. Similarly, existing native minorities (Black, Hispanic, and Indian) show much variation, yet economic need is clear.

There is some indication in our individualistic society that, as with the progress of other immigrants, individuals rather than ethnic groups are advancing. Whether it was the Irish, the Italians, or the Jews, the system has always favored the "best and the brightest" individuals. In contrast to Europe, American politics are typically not class politics. We tend to favor individual rather than collective solutions to social problems. What seems to be developing, then, is a small middle class among the newer ethnics. This middle class tends to be employed by the public sector (Moore, 1982). Depending on the perspective, the middle class has the function of helping or controlling lower-class ethnics.

SOCIAL WELFARE AND THE NEW IMMIGRANTS

The migration of the new immigrants to the city is a relatively new phenomenon. While social welfare researchers are beginning to gather information on their social problems and their income and labor force participation rates, reltively little is known about the participation in social welfare programs of the new immigrants.

The prevailing impression is that not much has been done to focus on the specific problems of the new immigrants. Until recently, they have been out of sight and out of mind. This denial of the problem stems in part from the nature of social problems in our society. Just because a problem exists and people need help does not mean that society will respond. For this to happen, a social problem must affect a large number of people and cause enough public concern so that action is possible (Gold and Scarpitti, 1967: 2).

Theodore Marmor (1973), using the example of national health insurance, has developed a model of how a social problem develops to the extent that broad societal action can be taken. Marmor feels that at the beginning of the policy process, efforts are made to study a problem. This study effort is carried out by university researchers and others in and outside government. This stage can be called the rational actor stage of policy development, since the main focus is upon rationally describing the social problem, the population affected, and the possible solutions. This stage is very important in publicizing the problem and gaining some support for action among the intellectual and governmental elite.

During the next stage, when it becomes clear that some action is possible, interest groups that can gain or lose from policy alternatives become involved in the policy development process. Under the organizational process stage, the skill and political power of various interest groups predominate in national policy determinations. Research and rationality which played some role in identifying and describing the parameters of the problem, are now secondary to the immediate vested interests of organizations in their efforts to determine policy choices.

Finally, individual actors and administrators become involved in order to influence and benefit from the final outcome of the policy. Many of the actors have different conceptions of the problems, investments in different outcomes, and different terms upon which they will compromise. For Marmor, during this phase, "where you stand depends upon where you sit." The resulting policy is a "collage of individual acts, outcomes of minor and major games, and foulups" (Marmor, 1973). While these phases in the policy development process follow sequentially, there often is overlap.

The situation of the newer urban immigrants is of a group whose social problems are just beginning to get attention. At this point, there does not seem to be major national concern about their difficulties. However, there are beginning efforts to discover and describe their conditions, and to develop rational approaches to social policy. There are an increasing number of books and articles on the subject. It is important to remember, however, that the field is still in the information gathering stage, that the problems of the new immigrants are not yet social problems that the general society feels should be given attention.

It seems evident, although further documentation is needed, that the existing social welfare system is increasingly being used by members of the newer ethnic groups. The income transfer system, which provides money to families with dependent children and to others who are unable to support themselves, has seen growing numbers of Spanish-speaking, Asian, and Native American clients. In some cities, Public Welfare Departments have translators to help their non-English speaking clients. Forms and announcements in these agencies are often in both

English and Spanish. This innovation attests to the growing Hispanic portion of the welfare population.

There have been major efforts by both public and private agencies in the relocation of Asian, Caribbean, and Central and South American refugees and immigrants to urban situations. Family counseling agencies, religious groups, and public welfare agencies all have made beginning efforts to help the newer urban immigrants to find food, shelter, and employment, and to make use of the existing social welfare system, if needed.

There have been some efforts by social agencies to help immigrant Asian children who are not part of families. Children without families have been resettled by two national, sectarian, private agencies. They are the U.S. Catholic Conference and the Lutheran Immigration and Refugee Service, both of which operate foster care programs. Children, many of whom were teenagers, have been placed in institutional settings and with families. There often have been problems because of cultural and language differences between children and their families. Teenagers who had been relatively independent before placement also have had difficulty adjusting to the controls imposed by family life. A study of the program recommended more group settings.

Minor children who came as "boat people" to America from Cuba and Haiti had very few social services. Children often were housed in barracks that were unsanitary and did not have structured recreational or educational programs. Eventually, these children were placed with foster families (Jenkins, 1983).

Efforts have been made in education to accommodate, somewhat, the special needs of the children of urban immigrants. Some public schools offer bilingual education programs. Various helping professionals, such as social workers, educational psychologists, and guidance counselors have been involved in providing supportive social services.

In the youth serving field, organizations such as the Boys Clubs, Girl Scouts, the YMCA and YWCA, and the settlement houses have had to alter their programs in order to serve teenagers from the families of the urban immigrants. In some cities, the United Way has funded special programs for the families of urban immigrants, especially for Hispanics.

Special governmental and private agencies have been established to work on specific problems of particular populations. Thus, in many cities, newer agencies have been developed to deal, for example, with the physical and mental health and social and cultural needs of urban Native Americans, Puerto Ricans, and Mexicans.

After resettlement, Asian refugees have tended to migrate to urban areas and form relatively self-sufficient cultural groups. Among Vietnamese refugees, there is a cohesive extended family network. "The Vietnamese self is defined less in terms of individual characteristics than of family roles and responsibilities. These mutual family tasks . . . promote a sense of interdependence, belonging, and support" (Timberlake and Cook, 1984: 8). Among the Vietnamese, at least, there is a self-help network, typical of rural communities, which provides somewhat for basic social welfare needs. According to Timberlake and Cook (1984: 8), "Theoretically, each extended family in Vietnam has its own health, mental health, and welfare system." However, outside help was needed to assist Southeast Asians in adjusting to the United States.

Resettlement programs developed by the federal government were important in connecting or introducing Asian Americans to private social welfare agencies. Grants went to private social welfare agencies for initial resettlement. These programs could provide job training, orientation, language instruction, physical and mental health services, family planning, and follow-up counseling. There was also a wide range of child welfare services authorized, including foster care, special educational services, health care, and other services for unaccompanied children until they became 18 years old (Wong, 1983).

In 1975, more than 500 mutual assistance associations were created within the Indochinese refugee community in the United States. These associations ranged from agencies the primary responsibility of which was the delivery of social services to cultural and fraternal voluntary associations. Some of these agencies were given some support from the Office of Refugee Resettlement (Bui, 1983).

By 1979, over 1,000 organizations familiar with the mental health needs of the Indochinese refugees were funded by the Indochinese refugee assistance program. These organizations included voluntary resettlement agencies, community mental health centers, Indochinese community organizations, and local mental health projects (Robinson, 1980).

The programs were developed relatively quickly and often were provided by agencies that had little appreciation of the cultures and lifestyles of the immigrants. Carl Wong (1983: 133) writes that "one problem has been that sponsors have tended to exert too much control over refugees, thereby imposing their norms and values on them. Refugees were discouraged by sponsors from applying for public assistance, which was not the 'American way.' Thus, many refugees were put in the position of taking low paying, entry level jobs—many of which were not commensurate with their skills and backgrounds."

There is some evidence that this trend is changing. One estimate was made recently that as many as 70 percent of the recent refugees, particularly from Southeast Asia, were receiving welfare of some kind in the greater Los Angeles area (Bluhm, 1983: 6).

Hispanics are the second largest and fastest growing minority in the United States. Mexicans, Puerto Ricans, and Cubans constitute the largest percentage of this group, although it includes refugees from Central and South America also (Querals, 1984).

While there are a few formal studies that deal with the use of social services by Hispanics, especially by undocumented workers, Hispanics increasingly are making use of traditional and nontraditional social services. Some Hispanics employ a *curandero* or a *curandera*, an indigenous folk health care expert, to cure them of mystical illness such as magical fright, anxiety, and the evil eye. When individuals with these difficulties are treated in a traditional Western medical setting, there is evidence that some of the problems and beliefs are cured, while others are not (Rivera et al., 1984). The family is central to Hispanics, and traditional sex-role distinctions tend to prevail. Social service workers who wish to treat Hispanics must understand their beliefs and culture (Laird, 1984).

While it is obvious that modest social welfare programs have been developed in many urban areas for some of the urban immigrants, the ultimate effect of these policies has not been determined clearly. Rather than helping, some authors feel that even the minimal social welfare services available are elements of social control. For example, Adalberto Aguirre, Jr. (1982) feels that social services have been used to divide various Hispanic groups, thus weakening their political power. The author goes on to suggest that because of language differences, Hispanics often do not totally understand their rights in the public welfare system, especially when services are provided by those from another culture. He feels that the major purpose of social services is to keep Hispanics dependent economically and weak politically.

While society has been less than generous in providing help for the newer urban immigrants, Asian Americans, Hispanics, and urban Native Americans have resisted using the few social services available to them. For example, Hispanic Americans underutilize available mental health services. Their utilization of mental health services is one-half or less of their representation in the general population. This disparity stems in part from cultural values that work against Hispanic Americans using services, the lack of accessible and relevant services, and the lack of workers who can bridge the cultural differences between pre-dominantly White agencies and their Hispanic clients (President's Commission on Mental Health, 1978a).

The same is true of urban Native Americans. While the Snyder Act authorizes services for Native Americans no matter where they work, many state and county welfare agencies have resisted serving Native Americans. The welfare agencies have felt that the Native Americans are being taken care of by the Indian Health Service. For their part, Native Americans have tended to rely for help on a strong family support system. To combat the resistance, many agencies serving urban Native Americans have developed advocacy programs aimed at convincing clients of their right to medical, financial, and mental health services and insisting that public agencies serve this population (President's Commission, 1978b).

Studies also show that Asian Americans consistently underutilize medical and social welfare services. They have a pattern of doctoring themselves and using folk remedies. "Asian and Pacific Americans generally have strong attitudes of shame, fear, guilt, ridicule, and rejection toward the mentally and emotionally disturbed member(s) of the family" (President's Commission, 1978b: 810). "Asian Americans . . . may perceive services such as counseling as shame-inducing processes and will undergo extreme stress when asking for or accepting help from anyone outside the family" (President's Commission 1978c: 791). Further, some of the counseling techniques are based on Western values that require a high tolerance for ambiguity, a great amount of verbal facility, and a willingness to reveal private thoughts and feelings. These values are not congruent with Asian American culture values.

THE CHANGING ROLE OF SOCIAL WELFARE SERVICES AND THE URBAN IMMIGRANTS

The future of social welfare services for urban immigrants must be viewed within the larger context of projections for social welfare generally. President Reagan has created massive changes within this area of public policy by reversing the federal government's historic role of attempting to limit dramatic social and economic inequalities. He has reduced significantly expenditures for social welfare services and decentralized their control and financing to the states and counties.

President Reagan has ensured the continuation of this trend through a tax policy of reducing the progressive federal income tax, while forcing states and cities to raise their more regressive taxes. By starting major new defense initiatives systems, the president has locked in increased expenditures for defense that many experts feel will last for at least the next ten years. Further, major interest groups that could be expected to support strongly increased social welfare appropriations, such as the labor unions, are having serious economic and political problems and cannot be expected to work toward policies that benefit nonunion members. On the other hand, interest groups representing the elderly have become increasingly powerful. As the total number of elderly grow as a percentage of the population, they can be expected to claim more of the social welfare dollar.

To a large extent, money available for social welfare policy is dependent on the growth of the gross national product. Aside from a major crisis, such as occurred in the 1930s and the 1960s, social welfare tends to grow as the economy grows and there is more money available. Whether the growth that occurred in the past will continue is questioned by many economists. Lester Thurow (1980), for example, feels that many problems facing the economy will limit growth and that there will be increasing competition for scarce governmental funds. Public policy is becoming like a zero-sum game, where a gain in one area means a loss in another area. Thurow (1980: 161) says that "we are entering a period of rising inequality where conventional income transfer programs will be incapable of preserving [even] the current degree of inequality."

In addition to following political and economic trends, social welfare policy reacts to demographic trends (Zald, 1977). The extreme cutbacks of the social welfare policies of the 1960s, along with the increasingly bleak employment outlook for many unskilled and low-skilled workers has resulted in an increase in the percentage of the population that is poor, emotionally ill, delinquent, and criminal. The success of the social policies of the 1930s and 1960s made possible the reduction of social welfare provisions during the 1970s and 1980s. We are now seeing a growing number of individuals who are desperate for help and who cannot be accommodated by the private sector. In addition, some of these individuals are becoming increasing active politically. In general, then, the conditions may be present for a modest general expansion in basic social welfare provisions, after the Reagan presidency. It also is possible that society could become more repressive and therefore foster social and political instability.

The situation of the newer urban ethnics is less certain. While they are increasing in numbers in urban areas, to date their growth has not seemed to be translated into widespread political victories. In the area of social welfare, they often have often resisted even the minimal help that society has made available. Cultural values characterized by a strong individualistic ethic have tended to restrict their support for and utilization of social welfare services.

However, as the new urban immigrants become more urbanized and discover that survival in the city is different from survival in a rural village and mountains, increased use of and need for social welfare provisions can be expected. Already, there are indications that the older immigrants, primarily the Hispanics and the Native Americans, are beginning to press for more adequate social welfare programs. It is probable that the Asian Americans, especially the second generation, also will participate in this effort. What has been called the "convergence of class and ethnicity" (Moore, 1981) will work strongly in this direction.

As several authors have observed, it is wrong to expect that the different newer urban immigrant groups will work together in a cohesive political force. Historically, American society has tended to create competition and division within lower classes based on ethnic factors (Greeley, 1969: 38-44). Even within ethnic groups, it is wrong to expect unanimity. As Joan Moore (1979: 78) writes, "Quite possibly, 'ethnic solidarity' may not be present in the Mexican American community now or ever, any more than it is present in Black communities."

To an extent, the society reacts, although often slowly, to pressure. In the absence of an organized effort to change social conditions, it is unlikely that there will be dramatic improvements in the condition of the newer urban immigrants. America historically has ignored its social problems until they have become serious and pervasive.

The first step in focusing attention upon the new urban immigrants includes studies of their characteristics and needs. More attention by "rational actors," to use Marmor's phrase, needs to be paid to urban immigrants. A survey of recent books and articles in professional journals indicates some minimal attention is being given to the new urban immigrants. Conceptually, additional work is needed in order for the helping professions to develop a theory and strategy for working with the newer urban immigrants. Older models and those from other countries may not be completely applicable to the newer immigrant but might offer a base upon which to build (Golan and Gruschka, 1971).

The old models of assimilation and acculturation are certainly not applicable today to the new urban immigrants. "Minority individuals must learn to function in two environments: their own culture and that of the mainstream society" (de Anda, 1984: 101). Bicultural social-ization, in contrast to assimilation or acculturation, seems to be a more realistic approach to the inclusion of the new immigrants into American society.

In the field of social welfare, there are significant gaps in our knowledge about what are priority needs among the various groups and how services can be designed to be responsive to particular cultural and class needs.

There is a major intellectual challenge to describe and analyze the new urban immigrants. Perhaps by focusing attention on the condition and needs of the urban immigrants, social science can have a positive impact on public policy.

9

Ethnicity, Mental Health, and the Urban Delivery System

WILLIAM T. LIU
ELENA S. H. YU

☐ THE MAIN PURPOSE of this chapter is to assess the relationship between ethnicity and mental health in American society. A related objective is to examine how the U.S. mental health care delivery system serves, or fails to serve, the needs of diverse ethnic groups in the country. To aid in accomplishing these two objectives, data are needed on the precise estimates of the *true* (as opposed to treated) prevalence of various types of mental disorders or some indicators of the mental health status and services utilization patterns of diverse ethnic groups vis-à-vis majority Americans. However, collection of interethnic psychiatric epidemiologic data by means of a standardized instrument did not begin until the end (circa 1980) of President Carter's administration and such data are available only for Whites, Blacks, and Mexican Americans in certain parts of the country (Regier et al., 1984), while similar data for other ethnic groups, such as American Indians, Eskimos, Asian Americans, and Pacific Islanders, are simply nonexistent. Therefore, one can approach the complex issues surrounding ethnicity and mental health only by converging findings from earlier (pre-1980) psychiatric epidemiologic research on the mainstream population with available census data and small-scale studies on a local level, so as to obtain an approximation of the size of high-risk groups among the new ethnics. In so doing, we are at the same time examining whether or not some of the

AUTHORS' NOTE: *The research described in this chapter was funded by NIMH's Center for Minority Group Mental Health Programs under grant 7 RO1 MH 36408. The authors are especially grateful to Ching-Fu Chang, Sociology Graduate Research Assistant, for tabulating the suicide statistics, and to Phyllis Flattery for her helpful comments and suggestions on an earlier draft. The timely assistance of Medy Masibay and Aiko Igarashi in typing the tables and the references is also very much appreciated.*

firmly established social risk factors uncovered in previous community surveys are in fact distributed in the same manner among the various racial and ethnic minorities as they are for the majority Americans. Furthermore, we seek to understand if, and why, the prevalence rates for some indicators of mental illness or psychological distress used in past community surveys are different for Blacks or Mexican Americans compared with White Americans.

In the following sections, we wish to make the conceptual distinction between race, ethnicity, and minorities even though we may sometimes use them interchangeably. Next, we will summarize epidemiologic studies on ethnicity and mental illness conducted between 1950 and 1980 in the United States, a time when various investigators were independently exploring different types of diagnostic instruments or screening scales. 1980 census data are introduced whenever applicable to throw light on the characteristics and distributions of selected ethnic groups in the United States and to identify the subpopulations at risk based on results of past psychiatric epidemiologic research. This is followed by a discussion of the de facto mental health delivery system and its limitations in dealing with the needs of non-Majority populations. The chapter ends with a discussion of the implications for change in the delivery of mental health care, and a set of research recommendations.

RACE, ETHNICITY, AND MINORITIES:
A POINT OF CLARIFICATION

Race, or the biological classification of *homo sapiens* by certain phenotypical and genotypical traits, is a construct that is not the focus of this chapter. The fact, however, that physical attributes can be *socially* defined is of utmost interest here. Physical traits, after all, have social and psychological ramifications. We note, for instance, that in the 1980 U.S. Census, Chinese, Japanese, and Koreans are identified as three distinct "racial groups" despite their belonging to the same Mongoloid race. Used in this sense, the census classification highlights the fact that these *newer* immigrants are neither Black nor White; different from one another even though they look alike. As such, the words "racial groups," as applied to Asians, actually approximates the meaning of ethnic groups.

Historically, ethnic groups in America have emerged as a result of one or several of the following processes: (1) migration; (2) consolidation of forces in the face of impending threat from an aggressor; (3)

annexation or changes in political boundary lines; and (4) schisms within the church. Hence, being ethnic presupposes something different from the mainstream culture. Nevertheless, to be "ethnic" is not to be equated with being "foreign," even though many of the sociological and historical studies of ethnic groups are focused on "foreigners," that is, the immigrants.

On the other hand, the words "minority groups," as first popularized by Louis Wirth, refer simply to victims of subordination or oppression regardless of their numerical size. Because the early sociological literature on immigrants in the 1920s tended to equate ethnicity with minority status, much emphasis was given to the existence of discrimination in America (Wirth, 1945: 364; Wagley and Harris, 1958: 4; Glazer and Moynihan, 1970). In reality, of course, the presence of distinct ethnic groups in any society has always created unequal opportunities for access to scarce resources. Even in the context of a pluralistic model, the concept of ethnicity has always been associated with discrimination.

Clearly, ethnicity and minority status are invariably overlapping concepts, and they have significant impact on the social, economic, physical, and mental health status of the people. As a relational concept, ethnicity cannot be understood without a reference group. Moreover, from a mental health perspective, ethnicity is a significant factor the greater the contrast between the parent and the mainstream culture. Thus, it may be hypothesized that the newer ethnics, who originate from a historical past that is distinctly different from the cultural heritage of the Western Christian tradition, are exposed to greater stress than the European immigrants who arrived decades earlier. The vast structural differences between Asia and America mean that much of the most basic economic and primary socioemotional needs of the new immigrants from Asia, especially Indochina and the Pacific Triangle, cannot be met by existing institutions of the host society. Instead, these recent newcomers are at once forced to adjust to an unfamiliar social setting by abandoning their traditional values, learning new social skills, and submitting to the painful process of acculturation in order to function properly in America (see Liu, 1979; Koh et al., 1984). This exposure to higher levels of stress, it is hypothesized, may produce higher rates of mental illness.

As victims of oppression and discrimination, ethnic minorities may have their peculiar, if not excess share, of stress burdens by virtue of their exposure to greater psychosocial risks that may lead to higher rates of mental illness. Nevertheless, they may also have certain adaptive lifestyles and unique social-psychological support systems derived from a cohesive social structure buttressed by ethnocultural values that moderate the impact of high levels of stress. The result, it is hypothesized, may be lower rates of mental illness.

Unfortunately, interesting though these hypotheses may be, empirical data on the distribution of stressors and resultant mental illness will not be available for comparison until results of the post-1980 Epidemiologic Catchment Area (ECA) studies are fully published.[1] Thus, for the time being, we need to review the other types of information available— however limited—in order to have a better understanding of the issues surrounding ethnicity and mental health.

DEMOGRAPHIC CHARACTERISTICS OF ETHNIC GROUPS

In terms of a number of commonly used demographic measures, the selected ethnic groups reported in the census are as diverse as can possibly be. For instance, while the sex ratios for Black and Japanese Americans are 89.6 and 84.5 males per 100 females, respectively, those for American Indian, White, and Filipino Americans are 97.5, 94.8, and 93.4 (in that order). Chinese and other ethnic groups, in contrast, continue to have slightly more males than females (102.3 and 102.8 respectively). Past immigration laws that restricted the family formation of Chinese in the United States since 1882 contributed to the persistent excess of males over females among Chinese.

The recency of the immigration influx from the Asia/Pacific Triangle is discernible when one examines the census nativity data. In the Asian and Pacific Islander group, 58.6 percent are foreign-born, compared to 28.6 percent of persons of Spanish origin, 4.9 percent of White Americans, 3.1 percent of Black Americans, and 2.5 of American Indians, Eskimos, and Aleuts.

One consequence of the high percentage of foreign-born is the transmission of a non-English language as the vehicle for communication within the family. Census estimates of language spoken at home by persons five years and over suggest that the Spanish, Chinese, Filipino, Korean, and Vietnamese languages are spoken at home by more than 90 percent of the persons who identify themselves as belonging to these respective ethnic groups (U.S. Bureau of the Census, 1983e: Table 256).

Furthermore, 90 percent or more of the Spanish and Asian American populations reside in urban areas throughout the United States, compared with 71 percent of White Americans, 85 percent of Black Americans, 39 percent of American Indians, 32 percent of Eskimos, and

53 percent of Aleuts. The fact that some subsets of these newer immigrants (such as the Hmong) have been air-freighted from a preliterate society to a modern high-technology society presents unique problems in cultural adaptation and social adjustments. Others, who moved from an industrial society to the metropolitan centers of the United States, are likely to be "unmeltable" because their physical attributes and linguistic handicaps preclude their "dissolution" in the American ethnic scenes no matter how assimilated they are to the mainstream culture. Without a doubt, the arrival of these modern immigrants contributes to the cosmopolitanization of American cities and creates pressure on the delivery of human and social services in ways previously unthought of in this country.

In the following sections, we will begin to ascertain the significance of ethnicity in mental illness by examining the best available data on the subject.

TREATED PREVALENCE RATES FOR ETHNIC GROUPS: CENSUS ESTIMATES

Based on estimates of a sample, the 1980 census provide some information on inmates of institutions for selected ethnic groups. Among the institutions relevant to mental health issues, only mental hospitals, correctional institutions, and homes for the aged have been tabulated separately.

The results show that insofar as inmates of mental hospitals (Table 9.1) and correctional institutions (Table 9.2) are concerned, there is a consistent excess of males over females across all ethnic groups. Among males, Black Americans and American Indians show an age-adjusted commitment rate to mental hospitals (2.44 and 2.40 per thousand, respectively) that is twice that found for White Americans (1.20 per thousand). Persons of Spanish origin and those who are identified as Asian/Pacific Americans have an age-adjusted commitment rate (1.00 and 0.45 per thousand, respectively) that is lower than that reported for White Americans.

Among females, only Black Americans have an age-adjusted commitment rate to mental hospitals that exceeds that found for White Americans (1.02 per thousand compared to 0.70 per thousand).

Likewise, the age-adjusted rates for inmates of correctional institutions (Table 9.2) indicate that, among males, the excess for Black Americans is about seven times that found for White Americans (15.13

TABLE 9.1 Estimates of Inmates of Mental Hospital per 1000 for Selected Racial Groups: United States, 1980

Age Group	White Male	White Female	Black Male	Black Female	Spanish Origin Male	Spanish Origin Female	American Indian, Aleut, & Eskimo Male	American Indian, Aleut, & Eskimo Female	Pacific Asian Male	Pacific Asian Female
All ages, crude	1.26	0.82	2.21	1.01	0.87	0.42	2.12	0.57	0.44	0.24
Age-adjusted[a]	1.20	0.70	2.44	1.02	1.00	0.49	2.40	0.63	0.45	0.24
0-4 years	0.02	0.01	0.03	0.03	0.01	0.01	0.11	0.01	—	—
5-14 years	0.26	0.13	0.46	0.17	0.26	0.10	0.24	0.08	0.06	0.13
15-24 years	1.17	0.54	2.02	0.72	0.90	0.29	1.98	0.72	0.60	0.21
25-34 years	1.53	0.74	3.83	1.19	1.30	0.47	3.50	0.71	0.72	0.21
35-44 years	1.53	0.90	3.31	1.28	1.26	0.77	3.82	0.67	0.31	0.19
45-54 years	1.63	1.07	3.46	1.41	1.28	0.66	2.99	1.30	0.51	0.53
55-64 years	1.86	1.21	3.26	1.78	1.60	1.01	3.94	0.89	0.63	0.54
65-74 years	1.98	1.41	3.10	2.19	1.82	0.99	2.74	0.55	0.69	0.14
75-84 years	2.44	1.97	4.12	3.17	1.75	1.20	3.36	1.58	1.08	0.19
85 years and over	3.69	2.82	5.92	4.36	2.69	2.58	10.20	—	0.54	1.90

SOURCE: U.S. Bureau of the Census (1983e: table 266).

a. Age-adjusted rates are what the commitment rates would be if age distributions were identical for all the ethnic groups under comparison. Age-adjustment is done by the direct method, using as the standard population the age distribution of the total population of the United States in 1940. Adjustment is based on ten age groups.

TABLE 9.2 Estimates of Inmates of Correctional Institutions per 1000 for Selected Racial Groups: United States, 1980

Age Group	White		Black		Spanish Origin		American Indian, Aleut, & Eskimo		Pacific Asian	
	Male	Female	Male	Female	Male	Female	Male	Female	Male	Female
All ages, crude	2.33	0.14	14.98	0.81	6.13	0.28	10.86	0.81	1.18	0.08
Age-adjusted[a]	2.33	0.14	15.13	0.81	5.94	0.27	10.52	0.77	1.11	0.08
0-4 years	—	—	—	—	—	—	—	—	—	—
5-14 years	0.06	0.02	0.17	0.03	0.07	—	0.10	—	—	0.04
15-24 years	5.09	0.28	26.65	1.36	10.99	0.52	21.75	1.72	2.85	0.20
25-34 years	4.89	0.30	40.31	2.29	13.87	0.66	21.75	1.75	2.32	0.14
35-44 years	2.71	0.15	18.26	0.92	8.18	0.38	12.59	0.78	0.71	0.07
45-54 years	1.43	0.10	7.41	0.33	3.53	0.11	7.46	0.30	0.56	—
55-64 years	0.56	0.04	3.03	0.15	1.35	0.03	4.19	0.16	0.47	0.04
65-74 years	0.23	0.04	1.29	0.09	0.48	0.02	1.26	0.17	0.44	—
75-84 years	0.08	0.10	0.14	0.11	—	—	—	—	—	—
85 years & over	0.06	0.23	0.47	0.06	—	—	—	1.04	—	—

SOURCE: U.S. Bureau of the Census (1983e: table 266).
a. See note a, Table 9.1.

per thousand to 2.33 per thousand); for Native Americans, about five times higher (10.52: 2.33); and for persons of Spanish origin, nearly three times over (5.94: 2.33).

Similar excess, though of a slightly smaller magnitude, is found in the female age-adjusted rates. Expressed in terms of ratio, the rate for commitment to correctional institutions found for Black American (0.81 per thousand) and American Indian (0.77 per thousand) women is about six times that found for White American women (0.14 per thousand). Females of Spanish origin show an excess ratio of only 1.93.

Census estimates of the number of inmates in homes for the aged throughout the United States point up a sex difference in the other direction (Table 9.3). That is, with the exception of American Indians, women 65 years and over across all ethnic groups have a somewhat higher institutionalization rate than men. In the two oldest age groups for which data are available (75 to 84 years and 85 years and over), White American males show a higher institutionalization rate than all other ethnic groups—again, with the exception of American Indians in the 75 to 84 age range. Among females, the excess in commitment rate for White Americans over other ethnic groups begins to surface at a much younger age, 65 to 74 years.

Interestingly enough, the presence of unusual number of persons under 45 years of age in the homes designed presumably for the aged is such that, for all non-White groups, there is a consistent excess in age-adjusted rates for males over females. While sampling errors might have accounted for this phenomenon, it is equally plausible that the constant pressure created by the Community Mental Health Center Act of 1963 might have brought about a transfer of patients from state hospitals to nursing homes because such transfer can be paid for by Medicare (Klerman, 1977; Clarke, 1979).

Obviously these data, however limited, give a national picture of the ethnic effect in institutionalized populations. They seem to suggest that where some kind of financial resources or knowledge about the availability of such resources are required for commitment, such as institutionalization in homes for the aged, the rate reported for Whites exceeds that found for all other ethnic groups. In contrast, the socially disadvantaged groups in American society, such as Black Americans, American Indians, and persons of Spanish origin, are more likely to have "behavioral" problems that result in commitment to correctional institutions.

The low institutionalization rate for Asian/Pacific Americans should be interpreted with caution, because more than half of them are

TABLE 9.3 Estimates of Inmates of Homes for the Aged, per 1000 for Selected Racial Groups: United States 1980

Age Group	White Male	White Female	Black Male	Black Female	Spanish Origin Male	Spanish Origin Female	American Indian, Aleut, & Eskimo Male	American Indian, Aleut, & Eskimo Female	Pacific Asian Male	Pacific Asian Female
All ages, crude	4.08	9.71	2.96	3.82	1.34	1.59	3.10	2.93	1.34	1.57
Age-adjusted[a]	2.72	3.72	3.14	2.78	1.98	1.78	4.21	2.99	1.49	1.38
0-4 years	0.03	0.05	0.04	0.02	0.03	0.01	0.00	0.00	0.09	0.06
5-14 years	0.09	0.06	0.16	0.11	0.07	0.05	0.51	0.40	0.02	0.02
15-24 years	0.24	0.16	0.38	0.22	0.24	0.10	0.46	0.49	0.14	0.06
25-34 years	0.51	0.36	0.79	0.40	0.39	0.19	1.30	0.42	0.19	0.21
35-44 years	0.88	0.72	1.40	0.74	0.56	0.30	1.41	1.16	0.22	0.35
45-54 years	1.85	1.80	3.45	2.00	1.38	0.81	3.21	1.71	1.00	0.57
55-64 years	4.38	4.66	6.78	5.01	3.54	2.13	9.12	5.66	2.06	1.33
65-74 years	13.39	16.89	16.81	14.89	10.40	8.84	21.56	13.76	8.28	6.61
75-84 years	46.40	80.17	41.37	47.62	33.06	39.85	68.94	50.24	28.09	27.43
85 years & over	157.31	264.31	84.23	134.66	90.96	107.63	104.17	149.69	78.44	112.86
65 years & over	32.07	62.51	27.56	34.44	21.58	24.79	39.47	35.46	17.44	21.69

SOURCE: U.S. Bureau of the Census (1983e: table 266).
a. See note a, Table 9.1.

immigrants. As such, selective factors that determine who immigrates to the United States militate against the likelihood of commitment to correctional institutions. Furthermore, some evidence is available that Asian/Pacific Americans tend to return to their parent countries for treatment of mental illness rather than face confinement in the United States, where the service providers are less familiar with their cultural conflicts and life stresses (Yeh et al., 1979).

LIMITATIONS OF INSTITUTIONALIZED DATA

Data based on treatment populations, however, do not provide useful information on the *true* prevalence of mental disorders in the community because selection factors determine who among those suffering from a particular type of psychiatric disorder actually receive treatment (Dohrenwend and Dohrenwend, 1969: 5-7). The problem is compounded by "the absence of standard case-finding techniques that can be used in a uniform and consistent fashion in population surveys to detect persons with mental disorders" (Kramer, 1976: 188). But the development of case-finding techniques depends on the existence of a consensus, which was lacking among members of the psychiatric profession as to what constitutes "psychopathology," "mental illness," or "psychiatric disorder." The publication in 1980 of the *Diagnostic and Statistical Manual, Version III* (or DSM-III, for short) by the American Psychiatric Association represents one of several attempts that started in the 1960s to develop a consistent definition of mental disorders. The release in 1981 of the Diagnostic Interview Schedule (called DIS), Version III (Robins et al., 1982) was in a sense part of NIMH's response to the President's Commission on Mental Health report (1978), which raised several questions about the prevalence rates for various types of mental disorders in the country and the methods used to obtain these rates, thereby leading to the development of a new case-finding technique for use in large-scale community surveys. To date, the reliability and validity of the DIS-Version III in community surveys both in the United States and across cultures have yet to be firmly established. Overseas, the DIS has been used in several independently organized community surveys in Taiwan, China, Korea, and Peru.[2] Similar epidemiologic surveys are being planned in Hong Kong and Manila. The results of these studies, when finally analyzed and fully published, should provide the best possible estimates to date on the *true* prevalence of mental disorders in several countries. Until then, we are forced to use other forms of available data to examine the ethnic factor in mental health and to obtain an appreciation of the size of high-risk populations in the United States.

TRUE PREVALENCE RATES: EPIDEMIOLOGIC FINDINGS

In general, psychiatric epidemiologic studies designed to provide data on the true prevalence of mental disorders in community settings fall into two broad categories: (1) those that yield information on types and rates of *mental illness* by means of *clinical* diagnoses based on the use of a diagnostic instrument or clinical judgments rendered by professionals, usually psychiatrists, on symptom data elicited through survey interviews, and (2) those that yield psychiatric *symptomatology* or psychological distress data but not clinical diagnoses. These studies employ actuarially calibrated *symptom scales* to screen persons who are "cases," that is, above a cut-off score that was either prespecified by previous validity studies with clinic patients or determined from newly studied samples of human subjects. Both sources of psychiatric epidemiologic data continue to suffer from serious methodological problems, but they provide the best information yet available for understanding the rates of mental disorders outside the institutionalized context.

PSYCHIATRIC EPIDEMIOLOGIC STUDIES USING CLINICAL JUDGMENTS

Between 1950 and 1980, only eleven psychiatric epidemiologic studies of adult populations using clinical judgments were conducted in the United States.[3] They represent a wide array of data collection procedures and case-identification methods. Data on non-Whites, though apparently available from some of these studies, were for the most part too small in number to analyze. Hence, they have been reported in only four studies. The first, conducted in Baltimore by Pasamanick et al. (1959), showed Whites to have far more mental illness than non-Whites (who were almost wholly Blacks).[4]

The second study, conducted in the 1970s, found no significant differences between the races in the lifetime or current rates for major or minor depression, nor in the lifetime rates for grief reactions or for cyclothymic and depressive personality (Weissman and Myers, 1978a, 1978b).

The third study focused on an American Indian tribe in a Pacific Northwest coastal village (Shore et al., 1973). Using a four-point scale from A to D, definite psychiatric disturbance (A rating) was assigned to 54 percent of the sample, and probable psychiatric disturbance (B rating) was found for an additional 15 percent of those studied. Alcoholism is the major psychiatric problem, if not the major health problem of this Indian village.

Data on yet another ethnic group, Mexican Americans, were collected in the mid-70s and just recently published (Vernon and Roberts, 1982). The results indicated that there were no strong ethnic differences in current rates of major or minor depression, although Whites had slightly lower rates than Blacks or Mexican Americans. Insofar as lifetime prevalence rates are concerned, Blacks had lower rates than Whites and Mexican Americans, whose rates were similar. Only in the bipolar II disorder (a diagnosis of low reliability and questionable validity) were rates for Blacks and Mexican Americans higher than those found for Whites (Vernon and Roberts, 1982: 51-52).

Since 1980, only one paper (Robins et al., 1984: 956), based on the use of the Diagnostic Interview Schedule (DIS) in five Epidemiologic Catchment Area (ECA) studies, has touched upon Black/White racial differences in true prevalence rates for mental disorders. The report, based on results from only three sites (New Haven, Baltimore, and St. Louis), stated that the difference in true prevalence rates between the two racial groups was quite modest and inconsistent across sites. Additional information on the socioeconomic breakdown of the two populations were not presented. Data on Spanish Americans are still being collected and only from one of the five ECA sites, while no psychiatric epidemiologic data are being planned for Native Americans or Asian/Pacific Americans.

SCREENING SCALE STUDIES USING CUT-OFF
SCORES TO IDENTIFY CASES

Although there have been numerous studies using screening scales to identify likely "cases" of mentally disordered persons, not all have collected information on race or ethnicity (e.g., Manis et al., 1964; Phillips, 1966). Of those that have, findings on non-White populations in the sample are not always analyzed separately. In the thirty-year period between 1950 and 1980, only fifteen screening scale studies are found in the literature that report findings on race or ethnic differences in "caseness." Insofar as Black-White differences are concerned, one study (Ilfeld, 1978) reported no differences in psychologic symptomatology while other studies have shown a weak but remarkably consistent result, despite the great disparity in survey methods and type of screening instruments used. Using either simple percentage distributions or zero-order correlations, Black Americans have been reported in these other studies to have had higher rates of depression or nonspecific psychological distress as measured by several screening scales such as the HOS, the CES-D, the Index of Psychological Well-Being, and similar types of instruments (Edgerton et al., 1970; Berkman, 1971; Warheit et al., 1973; Schwab et al., 1973; Warheit et al., 1975; Comstock and Helsing, 1976; Neff and Husaini, 1980; Roberts, Stevenson, and Breslow, 1981). However, where controls have been introduced, the race or ethnic differences almost always fail to reach statistical significance (Yancey et al., 1972; Warheit et al., 1973; Warheit et al., 1975; Comstock and Helsing, 1976; Carr and Krause, 1978; Neff and Husaini, 1980; Bell et al., 1981; Roberts et al., 1981).

Sex and *socioeconomic status* are by far the most significant factors in explaining the differences in crude rates observed for White and Black Americans. There are but only three exceptions (Berkman, 1971; Warheit et al., 1975; and Eaton and Kessler, 1981). In a nutshell, these exceptions are: (1) Blacks continued to score significantly higher than White Americans in the Phobia Scale used by Warheit et al. (1975), despite the presence of such controls as age, sex, and socioeconomic status; (2) income adjustment reduced but did not eliminate Black-White differences on the Index of Psychological Well-Being constructed from the Alameda County Study; and (3) perhaps the most intriguing, a greater proportion of Blacks are depressed at the extremes of the CES-D distribution (contrary to earlier findings by others). Furthermore, Blacks below the poverty level have the highest adjusted rate for depression of any group (Eaton and Kessler, 1981). These findings strongly suggest that Blacks suffer disadvantages not totally accounted for by differences in education, income, and other social class variables.

Compared to research on Black Americans, screening scale studies containing community samples of Spanish-speaking persons are extremely limited and the findings less consistent. For instance, Quesada et al. (1978) reported that Black females had higher scores (mean = 41.41) on the Zung Depression Scale than Mexican American females (mean = 37.82). Moreover, in their sample a larger proportion of the Black women (39 percent) than Mexican American women (25 percent) scored above the cut-off score for caseness on the Zung Scale. Antunes et al. (1974) found that Mexican American and Black respondents scored significantly lower than Anglos in mean number of symptomatic responses on the Langner 22-item Screening Scale. The Mexican American score was intermediate to the other groups, with Blacks showing the smallest mean number of symptoms. These same patterns held when socioeconomic status was controlled.

More recently, Roberts (1980) compared Mexican Americans, Blacks, and Anglos from two sample surveys in Alameda County, California, on several measures of psychological distress (such as satisfaction with leisure, marriage, job; reported happiness; positive and negative affect; reported episodes of emotional problems and chronic nervous trouble). He found that there were essentially no differences among the ethnic groups in either sample in adjusted or crude rates for emotional or mental illness or negative affect. There were, however, significant ethnic differences in satisfaction with leisure and with marriage, and positive affect, even after controlling for the effects of six covariates (age, sex, marital status, education, income, and physical health status); in no case are these rates significantly lower for Chicanos than for other groups.

With regard to reported happiness, adjustment for the effects of the covariates increased the difference between the Anglos and the Chicanos. The latter showing greater prevalence of unhappiness than the former, with Blacks showing the highest level of unhappiness. In yet another paper, Roberts (1981) reported that the mean depression symptom scores of Mexican American respondents were significantly higher than those of Anglos in the same samples taken from Alameda County. Adjustment for the effects of the six covariates did not eliminate the differences. Thus the results of these studies do not corroborate the earlier findings reported by Antunes et al. (1974) and Quesada et al. (1978), although Roberts's recent findings are corroborated by Frerichs et al.'s study (1981) in Los Angeles. There, using the CES-D cut-off of 16 points or greater to define a depressed case, the investigators found that Hispanics have the highest prevalence rate for depression (27.4 percent), White Americans the lowest (15.6 percent), with Blacks and others at an intermediate level (21.8 and 21.2 percent, respectively). After controlling for the effects of selected demographic and socioeconomic variables, neither race nor ethnicity were significantly related to the presence of depression. Obviously, more research is needed to clarify the issue.

Psychiatric epidemiologic studies, using clinical assessment tools, have yet to be conducted on other ethnic groups such as Asian/Pacific Islanders. One study, using a rating scale for depression—the CES-D, was published recently (Kuo, 1984), although the time when the data were collected was not reported. The results indicated that the mean CES-D score for the Asian American samples is higher than the mean of the White samples reported in other studies (e.g., Radloff, 1977; Frerichs et al., 1981). A greater proportion of the Asian samples (19.1 percent) had a depression score of 16 or above, which is higher than the rate found for White but lower than those reported for Hispanic and Black samples in Frerichs et al.'s study (1981). Among Asian Americans, statistically significant differences remain between Koreans, Filipinos, Japanese, and Chinese, even after holding constant several demographic variables (sex, marital status, age, nativity). Kuo's study remains to be replicated in other locales using a more rigorous method than what he had employed for identifying and sampling Asian ethnics.

In the absence of other types of psychiatric epidemiologic data, we will now examine the national mortality statistics in order to arrive at a more complete picture of the mental health status of several ethnic groups.

SUICIDE STATISTICS

Data extracted from death certificates submitted by each of the fifty states to the National Center for Health Statistics (NCHS) offer a rich but untapped source of information on death by suicide for Whites, Blacks, American Indians, Eskimos, Aleuts, and certain Asian American subgroups.

Table 9.4 shows the average annual age-sex-specific and age-adjusted death rates for six ethnic groups. Concentrating first on White/ Black differences, we note that both the age-adjusted and the age-sex-specific suicide rates for Black Americans are considerably lower than those for White Americans. Among the other ethnic groups examined, American Indians, Eskimos, and Aleuts have the highest suicide rates (14.23 per 100,000 population). In particular men in the 15 to 44 age range and women 15 to 34 years of age have the highest suicide rates of all when compared to other groups.

On the other hand, the suicide statistics for Chinese exhibit an interesting pattern. If one focuses only on the age-adjusted summary figure, Whites have a higher suicide rate (12.54 per 100,000 population) than Chinese (7.97). However, an examination of the age-sex specific rates indicates that in every age group starting from 45 years, Chinese women manifest a higher suicide rate than White women. In the oldest age group, the rate is 49.93 per 100,000 Chinese women compared to 4.92 for White women. In contrast, the Chinese male suicide rate exceeds the White male rate only in the 85 and over age group, a pattern

TABLE 9.4 Average Annual Age-Specific[a] and Age-Adjusted (1940 U.S. Standard) Death Rates[b] for Suicide, per 100,000 Population, for Specified Race: United States, 1980

Age Group	White			Black			American Indian, Eskimo, & Aleut		
	Total	Male	Female	Total	Male	Female	Total	Male	Female
All ages, crude	13.31	20.57	6.43	6.39	10.76	2.47	14.12	24.05	4.44
Age-adjusted	12.54	19.41	6.20	6.75	11.63	2.60	14.23	24.16	4.59
5-14 years	0.52	0.75	0.28	0.16	0.23	0.09	0.46	0.68	0.23
15-24 years	13.55	21.91	5.00	7.52	12.56	2.71	28.18	49.01	7.04
25-34 years	17.48	26.99	7.98	13.29	23.10	4.78	24.83	41.23	9.09
35-44 years	17.03	24.27	9.93	9.68	16.14	4.32	18.46	31.41	6.15
45-54 years	17.69	24.55	11.18	7.22	12.71	2.74	11.69	17.70	6.10
55-64 years	17.54	26.52	9.59	7.20	12.37	3.03	10.06	20.29	0.83
65-74 years	18.28	32.41	7.45	6.24	11.42	2.45	3.58	6.38	1.30
75-84 years	20.91	46.18	6.03	6.24	13.47	1.67	3.86	3.06	4.44
85 years & over	19.45	53.28	4.92	5.87	14.73	0.94	—	—	—

Age Group	Chinese			Japanese			Filipino		
	Total	Male	Female	Total	Male	Female	Total	Male	Female
All ages, crude	8.27	8.26	8.28	9.08	12.57	6.14	3.61	5.34	2.00
Age-adjusted	7.97	7.93	8.08	7.84	11.08	5.00	3.71	5.43	2.00
5-14 years	0.30	—	0.61	0.86	1.69	—	—	—	—
15-24 years	6.39	8.07	4.65	9.41	14.09	4.52	4.84	7.67	2.11
25-34 years	7.13	8.59	5.72	12.18	16.72	7.82	4.38	7.03	2.50
35-44 years	9.01	8.94	9.09	9.10	12.68	6.39	4.19	5.63	3.03
45-54 years	12.28	10.77	13.89	8.75	9.81	8.22	4.24	5.71	2.97
55-64 years	12.34	9.37	15.52	9.93	12.38	7.78	4.71	7.27	2.76
65-74 years	24.35	25.85	22.61	6.61	11.17	2.17	7.25	8.75	4.53
75-84 years	33.51	21.82	44.32	25.01	39.56	15.75	11.72	15.87	—
85 years & over	56.13	64.10	49.93	62.59	139.76	19.50	39.78	55.14	—

SOURCE: Division of Vital Statistics, National Center for Health Statistics, unpublished data calculated by the authors.

a. The numerator consists of 1979-1981 cumulative number of deaths; the denominator is based on the total enumerated of the 1980 U.S. Census.

b. Excludes deaths of nonresidents of the United States.

similar to the contrast between Japanese and White males. Japanese men have nearly as high an age-adjusted suicide rate as Black men, although the rate is still low compared to Whites. Japanese women, however, have a much lower suicide rate than Japanese men across age groups, which is lower than white Women except in the two oldest age groups, 75 to 84 years and 85 and over, suggesting the presence of a cohort effect. Finally, the Filipinos by and large have lower suicide rates than Whites with but one exception: Men in the oldest age group (85 and over) have slightly more deaths due to suicide than the comparable group among Whites. The reasons for such a dramatic contrast in the suicide statistics by ethnicity are unclear since systematic research using 1980 national mortality file at NCHS have yet to be conducted.

TRUE PREVALENCE OR TOTAL PREVALENCE?

The brief review of the literature in the previous pages demonstrates the complexity of the issues surrounding ethnicity and mental health. The dearth of psychiatric epidemiologic research on non-White groups and the virtual impossibility of disaggregating certain subgroups reported in published data (such as "persons of Spanish origin" in the census tabulations) makes it enormously difficult to make comparisons with or draw firm conclusions about specific ethnic groups (such as Mexican Americans) based on local-area studies. Nonetheless, the limitations of relying solely on either treated prevalence rates or "true" prevalence studies becomes abundantly clear as we compare the results of three sets of epidemiologic data salient to mental health research: institutionalized data, community survey findings, and suicide rates.

The caveat that institutionalized data present a distortion of the *real* prevalence rate for diverse ethnic groups must now be supplemented with another word of caution. Rates obtained from community studies on the distribution of mental disorders can be just as distorted because they count only the survivors and overlook those who might have committed suicide as a result of mental illness. While the specific cause of suicide cannot be known from death certificate data, and the association between mental illness (or poor mental health) and death by suicide remains to be established, there is sufficient evidence to be less confident that results obtained from "true" prevalence studies are true reflections of the *real* situation. Common sense would indicate that the overcommitment of Black Americans to both mental hospitals and correctional institutions *ipso facto* removes from the community a significant portion of the population *at risk* of mental disease, thereby depressing the upper-bound of the "true" prevalence rate obtainable from community surveys. Furthermore, both the Spanish-speaking and Asian/Pacific American groups contain a large proportion of foreign-born individuals. This suggest that selective migration factors may operate to make segments of these ethnic groups *constitutionally* different from, if not psychologically stronger than, what one might find had these groups not been largely immigrants. Furthermore, attachment to home countries among the immigrant groups provides the mentally disordered persons an alternative source of treatment (commitment to institutions in the home country), which lowers the treatment rate in the host society. Although precise figures on the extent to which this may have occurred is lacking, we cannot dismiss its potential significance in depressing the true rate of mental disorders for specific ethnic groups in the United States.

MINORITY STATUS, SOCIAL CLASS, AND CULTURAL HERITAGE

At this point in our review of the literature, the issue of ethnic group membership and mental health is not clear because social class, minority status, and cultural heritage are confounded. The hypothesis that ethnic minorities experience more distress than majority Americans has been broken into three separate arguments by Mirowsky and Ross (1980). We may call the first, the *racism* or minority status argument which was widely echoed during the civil rights movement in the 1960s. It states that ethnic group members, being minorities in the United States, suffer from prejudice and discrimination such that given the same education, minorities tend to hold less prestigious jobs than nonminorities. Furthermore, at every level of job prestige, minorities are paid less than nonminorities, and at every level of income, they may not be able to live where the nonminorities live or hold the same power positions in the community as the majority Americans do (Siegel, 1965; Miller, 1966; Duncan, 1969; Stolzenberg, 1975). In a presumably universalistic society oriented toward upward mobility, the downward pressure resulting from institutionalized racism may create a disjunction between expectations and achievement (Merton, 1957), which leads to feelings of lack of control over one's life, lowered self-esteem, and most likely higher rates of mental illness or psychological distress (McCarthy and Yancey, 1971).

A second argument, strongly suggested by the literature reviewed earlier, is that it is not racism associated with the minority status per se that generates psychological distress but low social class, to which most minorities belong, that leads to poor mental health. When socio-economic status is controlled for, differences between White and other minority groups tend to diminish significantly.

The third line of thinking posits that it is the distinctive beliefs, values, and behaviors of the ethnic minorities that are different from the mainstream culture, which produce a sense of alienation and threaten their mental health. The greater the culture gap between the majority group and the minority societies, the more strenuous the process of adaptation and coping strategies for ethnic minorities. Thus, this argument would posit that within the same ethnic community, the foreign-born immigrants have higher levels of stress and distress than native Americans.

In order to test these three hypotheses, the same data on diverse ethnic groups—selected to bring out the contrast between Western and Asian cultures—are needed, and socioeconomic status in the different groups must be controlled for in order to determine the effect of minority status on psychological well-being. Such a study has yet to be conducted.

Albeit the preponderance of evidence from community studies fails to show a higher rate of mental disorders or psychiatric symptomatology for Blacks compared to Whites, and despite the less-consistent findings reported for White-Spanish differences in psychiatric morbidity, sex and socioeconomic status still stand out as the two most critical variables in explaining the observed ethnic differences in existing studies. In the absence of empirical data, we are thus forced to accept for the time being the importance of socioeconomic status either as an independent variable in and of itself or as an exacerbating factor in predicting poor mental health, leaving aside the more significant issues of the independent effects of racism and cultural heritage in maintaining a person's psychological well-being. Under these circumstances, it appears that research on ethnicity and mental health would be better served if data are available to determine the proportional distribution of low socioeconomic status among the different ethnic groups.

IDENTIFYING HIGH-RISK GROUPS FROM CENSUS DATA

Table 9.5 show the number and percent below poverty level by ethnic groups based on the 1980 census. Compared to the 9.4 percent of Whites

TABLE 9.5   Persons Below Poverty Level in 1979
for Selected Racial Groups: United States, 1980

Selected Racial Groups	Total[a]	1979 Income Below Poverty Level	
		Number	Percentage
United States, total	220,845,766	27,392,580	12.5
65 years and over	24,154,364	3,581,729	14.8
female	113,907,899	15,756,126	13.8
White, total	184,466,900	17,331,671	9.4
65 years and over	21,691,260	2,774,505	12.8
female	94,757,233	10,018,588	10.6
Black, total	25,622,675	7,648,604	29.9
65 years and over	1,988,887	700,589	35.2
female	13,695,793	4,452,271	32.5
American Indian, Eskimo and Aleut, total	1,484,059	408,067	27.5
65 years and over	76,259	24,458	32.1
female	759,334	222,901	29.4
Asian/Pacific Islander, total	3,643,966	475,677	13.1
65 years and over	216,382	31,287	14.5
Female	1,895,390	244,395	12.9
Spanish origin, total	14,339,387	3,371,134	23.5
65 years and over	654,740	167,770	25.6
female	7,246,698	1,826,880	25.2

SOURCE: U.S. Bureau of the (1983e: table 304).
NOTE: Data are estimates based on a sample.
a. Exclude inmates of institutions, persons in military group quarters and in college dormitories, and unrelated individuals under fifteen years.

whose income of 1979 were below the poverty level, Blacks had 29.9 percent, American Indians, Eskimos, and Aleuts 27.5 percent, persons of Spanish origin 23.5 percent, and Asian/Pacific Americans 13.1 percent. The data suggest that while existing epidemiologic studies may fail to show a significant difference in psychiatric morbidity between different ethnic groups once controls for other variables are introduced, the disproportional distribution of certain covariates by ethnicity may be an important risk factor in mental illness.

Moreover, while the immigration factor creates a lopsided sex ratio for certain new immigrant groups that makes the ascertainment of gender distribution between ethnic groups uninformative with respect to psychiatric morbidity risks, the combination of being female and poor constitutes an added risk factor that should not be taken lightly if past epidemiologic findings are to be used as guidelines for future research on mental health and social policy. Table 9.5 indicates that while only 10.6 percent of White females reported income below poverty, 32.5 percent of Black females fell into that category, followed by 29.4 percent for American Indians, Eskimos, and Aleuts, 25.2 percent for persons of Spanish origin, and 12.9 for Asians and Pacific Islanders. In short, the percentage of women whose income was below the poverty level in 1979 is much higher for *every* ethnic group compared to Whites.

Furthermore, being old and poor are added risks that are disproportionally distributed between the different ethnic groups, with Whites showing the lowest rate (12.8 percent) compared to 35.2 percent for Blacks, 32.1 percent for American Indians, Eskimos, and Aleuts, 25.6 percent for persons of Spanish origin, and 14.5 percent for Asians and Pacific Islanders.

Table 9.6 shows the variation in poverty level between native and foreign-borns by year of immigration and country of birth. In the native category, only 9.4 percent of the families, 12.2 percent of persons, and 24.7 percent of unrelated individuals sharing a habitat, had incomes in 1979 below the poverty level, while in the foreign-born group, 20.7 percent of the families that immigrated between 1970 and 1980, 23 percent of persons, and 40.7 percent of unrelated individuals sharing a dwelling unit, fall into that category. It is interesting to note that the corresponding figure for those families, persons, and unrelated individuals who immigrated *before* 1970 is only 8.7 percent, 10.7 percent, and 24.3 percent, respectively. Recency of immigration is thus a critical factor in accounting for the high percentage of families living below poverty level.

Among the foreign-born families, persons, and unrelated individuals who immigrated during the last decade, those from Europe as a whole had a lower percentage below poverty (10.1, 12.4, and 30.1 percent, respectively) compared to those from Asia (19.2, 21.4, and 42.8 percent, respectively), Central America (percentage figure cannot be disaggregated from the published census statistics for "North and Central America" but can be surmised by comparing the figure for Canada relative to the other countries), South America (18.4, 19.5, and 39.8 percent, respectively), and Africa (21.1, 26.1, and 45.7 percent, respectively).

TABLE 9.6 Income in 1979 Below Poverty Level by Nativity, Year of Immigration, and Country of Birth: 1980 Census[a]

Nativity, Year of Immigration, and Country of Birth	Families		Unrelated Individuals		Persons	
	Number	Percentage	Number	Percentage	Number	Percentage
United States, total	5,760,215	9.6	8,860,582	25.1	27,392,580	12.4
Native	5,143,326	9.4	6,207,146	24.7	25,250,116	12.2
Foreign-born, immigrated 1970-1980	260,256	20.7	281,641	40.7	1,253,885	23.0
Europe	18,698	10.1	26,208	30.1	90,560	12.4
Asia	73,993	19.2	85,379	42.8	368,425	21.4
China	7,160	17.8	5,681	38.8	25,644	19.1
India	3,176	6.7	3,637	25.2	12,914	8.3
Korea	7,501	15.4	5,137	39.2	31,446	13.1
Philippines	3,848	6.2	6,440	22.8	21,258	6.8
Vietnam, 1975-1980	15,163	38.1	10,249	52.9	84,267	41.0
Vietnam, 1970-1974	427	19.1	392	28.7	2,027	12.4
North and Central America	124,932	25.1	106,370	42.7	574,101	27.0
Canada	2,141	8.4	5,384	30.6	13,960	11.4
Cuba	8,975	20.5	7,107	50.3	34,078	21.0
Dominican Republic	8,757	36.1	3,544	42.7	30,920	32.5
Haiti	4,017	25.6	3,542	45.0	15,105	25.7
Jamaica	4,193	15.1	4,513	32.5	18,394	16.3
Mexico	81,370	28.2	61,973	44.6	388,332	30.9
South America	13,671	18.4	16,195	39.8	60,384	19.5
Africa	7,104	21.1	13,356	45.7	32,459	26.1
Foreign-born, immigrated before 1970	266,633	8.7	371,795	24.3	888,579	10.7

SOURCE: U.S. Bureau of the Census (1983e).
a. Data are estimates based on a sample.

233

Within the new Asian immigrant groups, the influx from China, Korea, and Vietnam was highest during the decade of the 1970s and these same groups showed the largest percentage of families, persons, and unrelated individuals with incomes in 1979 below the poverty level (17.8 percent for immigrant families from China, 15.4 percent for those from Korea, and 38.1 percent for those who arrived from Vietnam after the fall of Saigon in 1975).

Similarly high, if not higher rates, are found for other physically visible immigrant groups, such as those families that immigrated from Cuba (20.5 percent), the Dominican Republic (36.1 percent), Haiti (25.6 percent), Jamaica (15.1 percent), Mexico (28.2 percent), South America (18.4 percent), and Africa (21.1 percent). Since the social and economic structures of these countries differ considerably from that of the United States, it is reasonable to postulate that unfamiliarity with the host environment will make it much more painful for these Asian and other physically visible immigrants to cope with their socioemotional needs and life stresses, compared to those originating from Europe and Canada. Unfortunately, no research has yet focused on these issues to identify the high risk groups for psychiatric morbidity by nativity and length of residence in the United States.

THE DE FACTO
U.S. MENTAL HEALTH SERVICES SYSTEM

The U.S. mental health services system is largely organized on a local basis and lacks uniform guidelines. Treatment services for the mentally ill are provided in both general health settings and in specialty settings for mental disorders, such as freestanding outpatient psychiatric clinics, general hospital psychiatric services, psychiatric hospitals, residential treatment centers for emotionally disturbed children, federally funded community mental health centers, and the offices of mental health professionals. There is no national- or state-level approach for coordinating health and mental health services setting relationships (Regier et al., 1978: 686).

In 1975, it was estimated that about 15 percent of the 213,032,000 total U.S. resident population have mental disorders during a one-year period and thus potentially require services. Of those estimated to be afflicted, 5 million or 15 percent received specialty mental health care services. An additional 3.4 percent received some diagnostic and treatment services in the General Hospital Inpatient/Nursing Home (GHI/NH) sector, while another 54.1 percent of those so afflicted in

1975 are estimated to have had contact with general medical professionals in Primary Care/Outpatient Medical (PC/OPM) sector (Regier et al., 1978). Yet another 6 percent of the mentally ill, or 2 million people, sought care in both health and mental health care sectors (National Center for Health Statistics, 1978). However, as many as 7 million people or 21.5 percent of the mentally ill were not treated or seen in either sector.

The 1979 national probability data on office visits within the coterminous United States made by ambulatory patients to nonfederally employed physicians who are principally engaged in office practice show consistent results (Yu and Cypress, 1982). Only 3.1 percent of the visits to office-based physicians in the total United States were made to the offices of psychiatrists. A breakdown of the data show that proportionally more White (3.2 percent) than Black (1.7 percent) or Asian/Pacific Americans (0.6 percent) visited a psychiatrist in 1979.

However, in 4.6 percent of the visits made by Whites, 3.1 percent of those made by Blacks, and 2.5 percent by Asian/Pacific Islanders, the *therapeutic* services rendered by the physician was psychotherapy or psychotherapeutic listening, thereby suggesting that practitioners outside of psychiatry were rendering psychiatric services of some sort. Mental status exams were rendered by the physicians as a *diagnostic* service in 1.5 percent of the office visits made by Whites, 1.3 percent of those made by Blacks, and 0.4 percent by Asian/Pacific Islanders—differences that were not statistically significant (Yu and Cypress, 1982).

In a recent publication based on the three Epidemiologic Catchment Area studies in New Haven, St. Louis, and Baltimore, Shapiro et al. (1984: 974) reported on ambulatory care utilization by the total adult population during six months prior to the interview. They found that close to three out of five persons visited a health or mental health care provider in that period; about 2.5 visits were made per person, and 4.5 visits per utilizer. However, between 6 and 7 percent of the adult population in the three cities made visit for a mental health problem to either a general medical care provider or to a mental health specialty source during the six-month period. So far, the ECA studies have not released data on utilization of mental health services by race or ethnicity.

Several local-area studies have indicated that Asian minorities are underserved and in general tend to underutilize mental health services, whether inpatient or outpatient (Kitano, 1969; State of Hawaii, 1970; Brown et al., 1973; Sue and Sue, 1974; Sue and McKinney, 1975; Sue and Morishima, 1982; Lin et al., 1982). This problem is shared by American Indians and Chicanos or Spanish-speaking minorities as well

(Karno and Edgerton, 1969; Sue and McKinney, 1975; Padilla, Ruiz, and Alvarez, 1975; Meketon, 1983). In contrast, Blacks—though underrepresented in visits to office-based psychiatrists—seem to be overrepresented in the use of community mental health centers.

TERRITORIAL BASIS FOR SERVICE DELIVERY

A basic cornerstone of the U.S. mental health service delivery system is the division of the nation into defined geographic zones called "catchment areas," where the Community Mental Health Centers (CMHCs) are to serve the needs of all the mentally ill persons in the given catchment. The Community Mental Health Center Act, instituted by Congress in 1963 in response to the deinstitutionalization movement that began in the 1950s, led to the establishment of over 640 centers throughout the country. Five basic services were specifically mandated for these CMHCs: (1) inpatient, (2) outpatient, (3) emergency, (4) partial hospitalization, and (5) consultation and education. The first four functions were designed to provide direct treatment to the sick, while the last one was viewed as fulfilling the prevention function for the community. The prevailing philosophy at that time was that by moving the locus of mental health care from state hospitals to the CMHCs, all individuals in the United States regardless of their social, economic, racial, or ethnic background, would be granted easy, if not unlimited, access and universal entitlement to mental health services (U.S. Congress, 1966; NIMH, 1971).Moreover, many of the undesirable consequences of long-term incarcerations and institutionalized care, assailed by numerous thinkers including sociologists such as Goffman (1961) and documented by psychiatric researchers such as Stanton and Schwartz (1954), would be averted.

Epidemiologic evidence indicates that the concept of community-based services has met with some success. First, national data (Witkin, 1980) demonstrate that, by 1977, long-term custodial care of patients in state hospitals had shifted to short-term inpatient and public outpatient treatment centers. As a result, state hospitals accounted for only 9 percent of all treatment episodes, down from 49 percent in 1955. Second, through the processes of *catchmenting* (targeting a well-defined geographical region) and *outreach* (actively seeking out untreated symptomatic persons), an enormous expansion in outpatient services was observed such that by 1977, the CMHCs accounted for the largest share of such services (32 percent of all treatment episodes). Third, dramatic changes in public attitudes toward mental health consultation was observed. The percentage of people seeking help for personal problems increased from 14 percent in 1957 to 26 percent in 1976 (Gurin et al., 1960; Veroff et al., 1981). Among the help-seekers, the proportion consulting mental health professionals has also increased markedly from 30 percent in 1957 to 49 percent in 1976. Fourth, the CMHCs have increased outpatient care for all lower-class groups except the elderly and some minorities (Tischler et al., 1972; Tischler et al., 1975a, 1975b; Babigian, 1977; Stern, 1977). The increase is especially noticeable among a subgroup of the lower class—women with less than high school education (see Gurin et al., 1960 and Veroff et al., 1981).

Nevertheless, although in aggregate non-Whites do use CMHCs at a higher rate than Whites (Redick, 1976), in reality the impact of these centers has not been evenly distributed among the different ethnic minority groups. Persons of Spanish origin and Asian Americans use mental health services far less than other groups (Karno and Edgerton, 1969; Willie et al., 1973; Sue and McKinney, 1975; Padilla et al., 1975; and Sue, 1977). An examination of the client and staffing loads in the CMHCs nationwide shows that, for Blacks, a significant association is found between services utilization and the availability of Black staff and professionals (Windle et al., 1979; Wu and Windle, 1980). Similarly, for American Indians, use of the CMHCs is significantly associated with access to Indian staff, but bears little relationship to whether Indian professionals are available. For Spanish minorities, on the other hand, use of services is associated with the availability of ethnic-specific professionals but not of Spanish staff, while no statistical association between use of services and staffing rate was found for Asian Americans. The precise causal relationships behind these measures of association cannot be determined from cross-sectional data. Suffice it to say that the desire for ethnic-specific delivery, conceived as one of the ideals during the height of the community mental health movement, has not been achieved.

Ominous signs of unanticipated consequences of the community mental health movement have also surfaced, namely: (1) organized community resistance to the mentally ill and the growing responsibility of the judicial system in determining who is insane (Whitmer, 1980; Borus, 1981); (2) the documented cases of patients now living in noninstitutionalized settings who failed to "graduate" from CMHCs or day hospitals even after such long periods of care as seven years (Scheper-Hughes, 1981); (3) the use, at greatly reduced costs, of nontraditional mental health workers who may have better rapport with lower-class patients, and a six-week inservice training, but in fact do not have adequate medical background or proper psychiatric residency training (Blum and Redlich, 1980; Fink and Weinstein, 1979); (4) the increasing reliance on psychotropic drugs to handle more patients, in order to exert social control on their behavior in the community and to show improved "productivity" or successful "access" over time and across centers; (5) the fiscal constraints and complex relation between the state and federal government with respect to Medicaid and Medicare reimbursement which transfer many patients, originally housed in state hospitals, into nursing homes without appropriate treatment programs or other alternative inpatient services (Klerman, 1977: 624; Clark, 1979); (6) the reduction, or elimination, from state budgets of discretionary programs such as vocational training for ex-patients and research on mental health that make assessment and long-term follow-up of the impact of deinstitutionalization even more difficult than before; and (7) the finding based on several studies that certain groups of lower-class patients (primarily minorities, the elderly, and chronic schizophrenics) continue to receive low levels of intervention (such as social support), are assigned to less experienced therapists, have higher rates of attrition, fewer treatment visits, less individual and family therapy, and more medication and "evaluations" (Mayo, 1974; Lorion, 1973, 1974, 1978; Sue, 1977; Flaherty and Meagher, 1980; Mollica and Redlich, 1980; Schubert and Miller, 1980; Mollica, 1980a, 1980b, 1983).

Perhaps one of the most curious aftermaths of the community mental health movement is the discovery that between 1970 and 1975 alone, the number of full-time equivalent psychiatrists in community mental health centers dropped from 6.8 to 4.3 per center, while the number of psychologists nearly doubled (from 4.9 to 8.5), and social workers increased from 9.7 to 12.2 full-time equivalents (NIMH, 1976; Winslow, 1979). No doubt, several factors—mostly social, political, and economic—contributed to the declining role of psychiatrists in CMHCs, which cannot be explained solely by the insufficient number of clinical psychiatrists (Redlich and Kellert, 1978). But, in the fifteen years since the passage of the Community Mental Health Act, the fact remains that the number of trained ethnic minority clinicians in the country has not increased significantly.

INSUFFICIENT ETHNIC PROVIDERS

A recent report by the General Accounting Office (GAO, 1983) points out that maldistribution of mental health professionals and the types of population served by these providers is a major problem in psychiatry *and* psychology. The most recent data show that as of September 1, 1984, Blacks represent only 1.8 percent of 41,460

TABLE 9.7   Ethnic Distribution of Psychiatrists in the United States, as of September 1, 1984, by APA[a] Membership

	Total		Percentage	
Selected Ethnic Groups	Number	Percent	APA Members	Non-APA Members
Total	41,460	100.0	72.4	27.6
White	35,140	84.7	73.5	26.5
Black	739	1.8	65.6	34.4
Native Americans	87	0.2	88.5	11.5
Mexicans	108	0.3	62.0	38.0
Puerto Ricans	175	0.4	56.0	44.0
Spanish	1,543	3.7	72.8	27.2
Asian Indians	1,806	4.4	66.3	33.7
Pilipinos	918	2.2	48.4	51.6
Chinese and Japanese	826	2.0	72.4	27.6
Other minorities	118	0.3	88.1	11.9

SOURCE: American Psychiatric Association, as calculated by the authors.
a. American Psychiatric Association.

psychiatrists in the country (see Table 9.7), a rate not much different from that found in 1967 when there were only 300 Black psychiatrists out of approximately 17,000 psychiatrists then in practice (Rosenfeld, 1976). The proportion of Native American psychiatrists, however, has increased from 0.1 percent in 1970 to 0.2 percent in 1984, just as the percentage of Spanish-surnamed psychiatrists has increased from 3.5 percent in 1970 to 4.4 percent (Mexican, Puerto Rican, and Spanish) in 1984. In contrast, the distributions of Asian American psychiatrists over time is more difficult to describe. While there were only 3.3 percent "Asian" psychiatrists in 1970 (Rosenfeld, 1976), the most recent tabulations from the American Psychiatric Association show 8.6 percent or 3,550 Asian psychiatrists, out of which 51 percent are Asian Indians and only 23 percent are either Chinese or Japanese. This representation is at odds with the distribution of Asian subgroups in the United States, where Chinese represent the largest Asian ethnic minority. It is not clear how many of these Chinese or Japanese psychiatrists actually speak a language other than English.

Not all psychiatrists are members of the American Psychiatric Association. Among Whites, only 73.4 percent are members of the professional association. Spanish and Chinese/Japanese have relatively similar figures (72.8 and 72.4 percent, respectively), compared to lower percentages for Asian Indians (66.3), Blacks (65.6), Mexicans (62.0), Puerto Ricans (56.0), and Filipinos (48.4).

TABLE 9.8 Ethnicity of Provider by Type of Provider and Degree Level[a]

| | Health Services Only | | | | Both Health and Educational Services | | | | Educational Services Only | | | |
| | Doctoral | | Master's | | Doctoral | | Master's | | Doctoral | | Master's | |
	N	%	N	%	N	%	N	%	N	%	N	%
Total	10,209	100.0	1,217	100.0	5,907	100.0	730	100.0	2,099	100.0	618	100.0
Ethnic Groups:												
White	9,810	96.1	1,179	96.9	5,631	95.3	706	96.7	2,011	95.8	573	92.0
Black	120	1.2	7	0.5	96	1.6	14	1.9	35	1.7	18	3.0
Hispanic	98	1.0	10	0.9	48	0.8	6	0.8	14	0.7	9	1.4
Asian	102	1.0	19	1.5	54	0.9	0	0.0	34	1.6	5	0.8
American Indian	19	0.2	0	0.0	9	0.2	1	0.1	3	0.1	2	0.3
Not specified	60	0.6	2	0.1	70	1.2	3	0.5	2	0.1	11	1.8

SOURCE: G.R. VandenBos and J. Stapp, "Service Providers in Psychology: Results of the 1978 APA Human Resources Survey," *The American Psychologist*, Vol. 36 (December, 1983), Table 1. Copyright 1983 by the American Psychological Association. Reprinted by permission of the authors.

a. Column percentages are given. The application of fractional weights and rounding may result in total percentages and Ns that differ slightly from the sums of subgroup percentages and Ns. In each case, Ns have been rounded to the nearest whole respondent.

Insofar as psychologists are concerned, the most recent survey conducted by the American Psychological Association (APA) shows that the majority (95.7 percent) of the psychological service providers are Whites (not of Hispanic origin). Among those psychologists who work in educational services, the figure is 95.1 percent (Table 9.8). Altogether, 95.8 percent of the psychologists who responded to the APA survey in 1982 either provided only health services alone or a combination of health and educational services (called "pooled health service" providers for short). Blacks (not of Hispanic origin) represent 1.3 percent of the pooled health service providers; Hispanics (regardless of race) 0.9 percent; Asians 1.0 percent, and American Indians 0.2 percent. Likewise, among psychologists who provide only education service to their clients, Blacks represent 2.0 percent, Hispanics 0.8 percent, Asians 1.4 percent, and American Indians 0.2 percent. These findings are nearly identical to those reported in an earlier survey of a similar nature (VandenBos et al., 1981), leading VandenBos and Stapp (1983: 1332) to conclude, "Although the number of service providers has increased, the percentage of minority service providers remains low."

In terms of psychiatric or psychological services rendered, both the 1979 National Ambulatory Medical Care Survey data analyzed by Yu and Cypress (1982) and the American Psychological Association data reported by VandenBos and Stapp (1983) are consistent in their conclusion that a large share of the psychiatric and psychological services to minority clients are currently being rendered by nonminority providers and by professionals who are not specially trained to deliver such services. This brings a number of worthy issues to our attention.

Language barriers. Considering that three-fifths (59 percent) of the Asian and Pacific Islanders and nearly a third (29 percent) of the various Spanish populations counted in the 1980 census are foreign-born, the claim made by minority spokespersons that language barriers constitute a real impediment in service delivery or provider-client communications cannot be dismissed (Bloombaum, Yamamoto, and Evans, 1968; Karno and Edgerton, 1969; Kline, 1969; Padilla et al., 1975; Quesada, 1976). This problem is especially serious for Asian/Pacific Islanders for whom, the U.S. Census estimates, a language other than English is spoken at home by nine out of ten persons five years or older.

To overcome the language barrier in service delivery, providers have resorted to the use of interpreters, not all of whom are trained in mental health. While they represent the best "stop-gap" measure in the short run to solve the shortage of manpower, the distorting effects in the long run of interpreter-mediated diagnostic evaluations is a serious threat to the quality of services rendered to non-English-speaking minorities. Albeit systematic research on this subject has yet to be conducted, the little evidence available from anecdotal cases suggests that the distortions that arise from the use of interpreters may sometimes substantially understate the degree of emotional disturbance endured by non-English-speaking persons seeking care (Sabin, 1975; Marcos, 1979).

Differences in cultural heritage. The demand for more bilingual staff in social service agencies and clinic settings overlooks the fact that being bilingual does not necessarily mean being bicultural. If bilingualism is difficult to assess, biculturalism is even more more difficult to determine, much less to impart in a programmatic fashion. After all, the root of the issues surrounding ethnicity and mental health lies in the ubiquity of culture. Thus, even within an ethnic population, differences in the cultural heritages of those who are foreign-born and those who are native can be just as critical a factor in the underutilization of mental health services as language barriers (Sanua, 1966; Yamamoto et al., 1968; Luborsky et al., 1971; Wolkon et al., 1973; Grantham, 1973; Halpern, 1973; Tseng, McDermott, and Maretzki, 1974; Valle and Fiester, 1976). In addition, intraethnic social class differences can be as problematic in help-seeking encounters as differences in cultural heritage (Carhuff and Pierce, 1967; Cobb, 1972; Cole and Pilisuk, 1976; Umbenhauer and Dewitte, 1978). Clinicians such as Yamamoto et al. (1968) and Abad et al. (1974) have indicated that the middle-class approach of psychotherapy where the patient is seen for fifty minutes once or twice a week, alone or in a group, tends to discourage lower-class patients from returning for services because they are more accustomed to personalized attention on a spontaneous basis. To complicate matters, the interaction between sex roles and cultural heritage can produce rather sharp contrasts in behavioral expectations that, when superimposed on a problem in communications, can easily lead to underutilization of mental health services.

In the current model of delivering mental health services to ethnic minorities, it is not always clear that service providers are able to distinguish between the patient's *cultural* style of coping and his or her *individual* style of dealing with life crises, for the acquisition of such insights have been left to the practitioner's haphazard discovery during practice instead of being systematically imparted in his professional training.

Use of alternative resources. Gurin et al. (1960) reported that most Americans seek help from clergymen and family physicians for emotional problems. Ethnic minorities, raised in indigenous cultures that differ from that of mainstream America, are not equally at ease in approaching such individuals for their family or personal problems. Instead, several authors have suggested that they tend to exhaust alternative sources of indigenous care before or even while seeking specialty mental health services (Padilla, Ruiz, and Alvarez, 1975; Comas-Diaz, 1981; Egawa and Tashima, 1982). Such use has been attributed by researchers to a number of factors, namely: (1) the degree of acculturation; (2) the high cost of conventional mental health care in the United States; (3) the lack of knowledge concerning the location and services offered by the de facto U.S. mental health services system; (4) inconvenient hours of operation and inflexible office procedures; and (5) different beliefs concerning the etiologies, diagnosis, and treatment of mental illness and what constitutes appropriate health promotion behavior. Therefore, it is important to keep in mind that the use of alternative sources of health care may *not* always be due to the ethnic patient's preference for these treatment modalities. Rather, they may be the last resort for disadvantaged ethnic groups who have no other recourse for their problems.

CONCLUSIONS AND RECOMMENDATIONS

There clearly is a difference in the mental health problems and adaptive skills required of the "old" immigrants, mainly German and Scandinavian settlers who arrived at a time when American society was still predominantly agrarian, compared to the "later" immigrants, such as Irish farmers and Polish peasants who found their way to the early industrial centers (Thomas and Znaniecki, 1958), and the "modern" immigrants who are transported by jet planes from preindustrial societies to metropolitan cities.

The arrival of Asian immigrants in the United States, however, was an event that was not anticipated by sociological theorists, and their mental health problems or adaptive behavior certainly cannot be adequately predicted by existing theoretical formulations in race and minority studies.

The concept of catchment areas, which form the ecological basis for mental health service delivery in the United States, would have effectively served the needs of ethnic minorities in America before World War II when Blacks were highly segregated and immigrants came from homogeneous backgrounds, arrived in identifiable waves, and settled in clustered pockets of the city. However, two major historical forces after World War II—the settlement of Black rural workers in the inner cities of metropolitan areas and the suburbanization of urban Blacks, as well as the influx of Asian and Spanish immigrants with diverse technical skills and education—have transformed the American urban scenes to the point that the ecological theory of cities developed by Park et al. (1925) and others can no longer be applied meaningfully. The new ethnics, who failed to "melt into the pot," also failed to lose their identity (or become "marginal") as had been predicted by theorists from the Chicago School. Community attachments were formed, instead, on the basis of a mixture of cultural, ideological, and interest proximities (Suttles, 1972). Consequently, although remnants of territorial segregation still exist, ethnic communities can no longer be easily identified by their territorially defined religious and service agencies. Rather, they are identified by political reformers, extended and outreaching cultural and employment services, ethnic and racial leaders, homeland governments, and philanthropic or civic leaders (see Kornblum, 1974; Quimby, 1976; Hoover, 1979; Kim, 1979). In the quest for community and neighborhood services and the right to determine their own lifestyles, the different ethnic elements have aligned themselves and merged political and business leaderships across ethnic lines.

As the delivery of mental health services became decentralized by the establishment of the CMHCs, control of the decision-making processes was sometimes taken over by politically ambitious community advocates, often without the effective participation of either the professional service providers or the sick who truly needed services. In more than a few instances, the continuing struggle among the various strata of mental health professionals and paraprofessionals for authority and control have weakened the position of all professionals. Because psychiatrists' services are expensive, few are hired by the CMHCs. Moreover, the position taken by psychiatrists trained to look at mental illness from a psychodynamic or medical perspective—that patients in some way contribute to their own distress—is often exploited by community activists to create an air of antiprofessionalism. In some cases, psychiatrists have been driven out of the centers (Fink and Weinstein, 1979).

The current model of community mental health service centers has been critically evaluated by Bachrach (1976) and others (Braun et al., 1981). The most vexing problem in evaluating CMHCs, of course, is the lack of consensus on the criteria of evaluation and on the objectives or priorities of the centers, in addition to the difficulty of defining the meaning of "community" itself. The failure to establish a viable and permanent mental health service facility for Indochinese refugees, for instance, lies precisely in the failure to identify a territorially stable Indochinese community in any city of the United States, despite the fact that about 50 percent of all Indochinese refugees are residing in California. This problem is less severe for Blacks, Native Americans, and some communities of Spanish persons, most of whom are still segregated and can be catchmented for services. It exists, however, for Asian ethnic groups such as East Indians, Koreans, Filipinos, and Chinese, who are increasingly choosing not to live in ethnic ghettos. Dispersed, these smaller groups are not drawn in sizable numbers in any national sample surveys on the physical or mental health of Americans (Yu, 1982; Yu, Drury, and Liu, 1982; Liu, 1985). Neither are they served by ethnic-specific professionals at the CMHCs (see Wu and Windle, 1980). Thus, both in research and in service delivery, these ethnic minorities do not exist in sufficient numbers within a geographic boundary (except for specialized communities such as Chinatowns) to merit serious attention. Despite growing interest and concern for minorities, they remained neglected.

It is unfortunate that the establishment of the CMHCs was not wedded to the provision of psychiatric training for ethnic minorities and that the politicization of decentralized mental health care has discouraged clinical psychiatrists from working CMHC settings. As a result, although such minorities as Blacks and American Indians are utilizing existing services more than other groups, the *quality* of care that they receive is of dubious value.

The creation of CMHCs, though beneficial in many ways, has also inadvertently diverted attention from the social origins of mental illness: poverty, racism, unemployment, and poor housing, among others. It is paradoxical that a movement that surged as a result of public awareness about the limitations of institutionalized care and the importance of *social* factors in restoring health would fail to advance the availability of psychosocial technology or expand the range of psychotherapeutic techniques to groups most deserving of such services. Instead, the CMHCs have come to depend increasingly upon psychotropic drugs to control the behavior of patients living in the community. As a result, adverse neurological complications of potent drugs are now becoming a serious problem among patients transferred out of mental hospitals. Meanwhile, the economic condition and ecological landscape of the community, wherein the social origins of mental illness lie, has for the most part remained unchanged. In many ways, deinstitutionalization has simply failed to reverse the social and occupational deficits accompanying mental illness—some ethnic groups being affected more acutely than others.

In the decades ahead, the mental health needs of various ethnic minorities would be better served if systematic efforts (including congressional appropriations) can be made to increase the supply of professionally trained ethnic service providers, especially psychiatrists andpsychologists, and a programmatic approach is made to broaden the ethnocultural horizons of the existing pool of mental health professionals. Additionally, our knowledge of the relationship between ethnicity and mental health would be greatly advanced if more research is conducted with the following aims: (1) to disentangle the effects of racism, social class, and cultural heritage on psychological distress and mental illness; (2) to understand the public attitudes toward mental health among the different ethnic groups and the relationship of such attitudes with use of services; (3) to determine the effectiveness or harmfulness of various types of intervention programs, both psychotherapeutic and biomedical; (4) to understand the decision criteria employed by hospital staff in determining whether or not to grant admission to voluntary applicants, to transfer patients into nursing homes, or refer them elsewhere. Knowledge of the sociodemographic profiles and subsequent fates of persons who are denied admissions is important to ameliorate the inequity of the present mode of mental health service delivery; and (5) to propose better ways to modify the social environments so as to prevent the rapid deterioration of patients released from state hospitals. Indeed, changes in the community structure of this country will have to made if institutionalized patients are to be truly taken care of outside the hospital setting.

NOTES

1. The recently published data, which appeared in the October 1984 issue of the *Archives of General Psychiatry,* gave only a summary of findings from samples of Black and White Americans in New Haven, Baltimore, and St. Louis. The Hispanic data are still being collected, and no psychiatric epidemiologic survey data are available for Native Americans or Asian/Pacific Islanders.

2. The study in Taiwan is conducted by Eng-Kung Yeh, M.D., and Hai-Gwo Hwu, M.D., of Taipei City Psychiatric Center and the Department of Psychiatry, National Taiwan University. The research in Mainland China is a collaborative project between the Shanghai Psychiatric Hospital and the Pacific/Asian American Mental Health Research Center (P/AAMHRC) of the United States. The Hong Kong Project is conducted under the leadership of Char-Nie Chen, M.D., Department of Psychiatry, Chinese University of Hong Kong. The survey in Korea is directed by Chung-Kyon Lee of the Department of Psychiatry, Seoul National University, in consultation with P/AAMHRC and with Joe Yamamoto, M.D. of the Neuropsychiatric Institute at UCLA. Two independent surveys

have been conducted in Peru using the DIS, one led by a team of Japanese psychiatrists in consultation with Joe Yamamoto and another headed by Filogmino Gaviria, M.D., of the Department of Psychiatry, University of Illinois at Chicago.

3. These studies are Eaton and Weil (1955), Pasamanick et al. (1959, 1962), Trussell et al. (1956), Dohrenwend et al. (1971), Srole et al. (1962), Cole et al. (1956), Schwab and Warheit (1972), Shore et al. (1973), Langner and Michael (1963), Weissman and Myers (1978a, 1978b), and Vernon and Roberts (1982).

4. The rate per 1000 persons for psychoses was 5.8 for White Americans, which was approximately nineteen times that found for Blacks (0.3). For psychoneuroses, the White rate (62.2 per 1,000) was more than twice that reported for Blacks (27.5). Likewise, autonomic and visceral disorders of psychophysiologic origin were almost three times as high for Whites (43.7) as for Blacks (17.7).

10

Rainbow's End:
From the Old to the New
Urban Ethnic Politics

STEVEN P. ERIE

☐ IN THE PAST TWENTY YEARS there have been dramatic population shifts in Sunbelt cities such as Los Angeles, Miami, and San Antonio. Cuban Americans now represent over one-half of Miami's residents, while Mexican Americans constitute over one-half of San Antonio's population. Los Angeles, however, serves as the nation's new Ellis Island. Since 1970 over 2 million foreign-born have settled in the metropolitan area. The "new" immigrants of predominantly Hispanic or Asian background have flocked to these cities as a result of changes in the nation's immigration laws, political and economic crises in the Third World, and new convergences in international capital and labor markets.

Hispanics and Asians represent the fourth large-scale wave of ethnic migrants to the big cities since the 1840s. Western European immigrants—primarily Irish and Germans—constituted the first massive wave from the 1840s to the 1870s, settling in the cities along the Atlantic seaboard and in the upper Midwest. Southern and Eastern European immigrants—Jews, Italians, Poles—formed the second wave of migrants, coming from the 1880s to the early 1920s, settling in many of the same northern cities as had the Western Europeans. In the post-World War II era, Blacks and Puerto Ricans constituted yet a third wave of migrants to the northern cities.

What urban political forms does the fourth-wave migration promise to take? How do these emerging forms compare with the brands of urban politics practiced by earlier ethnic arrivals? This essay compares and contrasts the old and new urban ethnic politics. Specifically, I examine the urban political experiences of four politically active groups:

AUTHOR'S NOTE: I wish to acknowledge the careful comments and criticism of Harold Brackman, Ellen Comisso, Victor Magagna, Val Melloff, and Joan and Burt Moore in the preparation of this chapter.

the Irish, Jews, Italians, and Blacks. Their patterns are compared with the political behavior of Mexican, Cuban, and Asian Americans.

Social scientists are only now beginning to shift their attention from the political behavior of established groups to the fourth-wave migrants. As they do so, they invariably compare the newer groups with the older. To the extent that the terms of the debate over the emerging character of Hispanic and Asian American urban political participation are influenced by our understanding of the past, we need to carefully separate historical fact from fiction.

This essay seeks to dispel two myths regarding the manner of previous ethnic political incorporation. The first concerns the White European immigrants; it may be termed the "rainbow" theory of the old-style urban machine. The theory refers to both the *players* and *prizes* of urban politics. In this view, urban machines actively worked to incorporate the European immigrants. Machines supposedly fashioned multiethnic "rainbow" coalitions, rewarding each group with jobs and services drawn from a sizable pot of municipal gold.

It is time to lay the rainbow legend to rest. In this essay I argue that throughout most of their history, the urban machines did *not* incorporate the European immigrants except for the Irish, who captured and defended the Democratic machines. Owing to the modest and fragile nature of the prizes of urban politics, the Irish could not readily translate political power into group economic advancement. Limited as these prizes were, the Irish jealously guarded them, parsimoniously accommodating the later-arriving Southern and Eastern Europeans. The newcomers struggled constantly with their Irish political overlords. The forms of anti-Irish insurgency were varied: third parties, reform movements, and revolts within the machines. For the Southern and Eastern Europeans, integration into the urban machines was a delayed, hard-won, and ultimately limited accomplishment.

The second myth concerns the manner of Black urban political incorporation. Urban machines did not accommodate Blacks. Moreover, Blacks were unable to carve out niches in the reformed bureaucracies that replaced the machines. By the 1970s and 1980s, however, local elected office represented an important mechanism of elite incorporation. Black mayors have been elected in Los Angeles, Chicago, Philadelphia, Detroit, Atlanta, Washington, D.C., Gary, Newark, and New Orleans. Blacks have also been elected in increasing numbers to city councils and school boards. Yet an exclusive focus on the urban electoral arena overlooks a prior and largely hidden mechanism of Black mass incorporation: the welfare-state programs of the New Deal and Great Society. The social programs of the 1960s generated service jobs for the new Black middle class. New Deal and Great Society welfare programs promoted the growing dependence of the Black poor upon public transfer payments. By the mid-1970s, Blacks were more economically integrated into the public sector than the Irish had been during the machine's heyday. In the following pages I highlight the degree to which urban Black politics has been shaped not by the vicissitudes of local candidates and issues but by the dynamics of the American welfare state.

THE IRISH:
BUILDING AND DEFENDING THE URBAN MACHINES

Between 1846 and 1870 nearly 3 million Irish immigrants came to the United States. Landing in the Eastern port cities of New York, Boston, and Philadelphia, the famine emigrants followed canal and railroad projects westward to the growing cities of upstate New York and the Great Lakes. By the 1880s, the Irish diaspora had dramatically altered the complexion of cities such as New York, Boston, Philadelphia, Chicago, San Francisco, Pittsburgh, Albany, Jersey City, and New Haven. First- and second-generation Irish constituted between 20 percent and 40 percent of the population of these cities.

The Celtic urban invasion soon took political form. By the late nineteenth century the machine had emerged as the dominant urban political institution; the Irish were its leading architects and practitioners. As a form of clientele politics, the party machine organized the electorate in order to control the tangible benefits of public office—patronage, services, contracts, and franchises. The machine employed these tangible divisible benefits in order to maintain power. Bosses purchased voter support with offers of public jobs and services rather than by appeals to class-based interests or to traditional loyalties (Banfield and Wilson, 1963: 115-117; Scott, 1969: 1143-1145; Wolfinger, 1972: 374-375).

With roots in the second or Jacksonian party system of the 1820s and 1830s, the full-fledged or mature urban machine did not emerge until the advanced stage of the third party system in the 1880s and early 1890s. By the 1880s the Irish controlled most of the Democratic machines being constructed in the Northern cities. Lamenting the "Irish conquest of our cities," Yankee John Paul Bobcock in 1894 furnished a roll call of the late nineteenth-century Irish bosses: John Kelly and Richard Croker in

New York City, Pat Maguire in Boston, Mike McDonald in Chicago, Chris Buckley in San Francisco, and "Little Bob" Davis in Jersey City (Bobcock, 1894: 186-195; Clark, 1975). In the twentieth century, Irish bosses would rule New York (Charles Francis Murphy), Chicago (Ed Kelly, Pat Nash, Richard Daley), Pittsburgh (David Lawrence), Jersey City (Frank Hague), Albany (Dan O'Connell), and Kansas City (Tom and Jim Pendergast).

The big-city machines run by the Irish are now in eclipse. Their welfare and employment functions have been supplanted by labor unions and government bureaucracies, their ethnic constituents have moved to the suburbs, and their control over patronage has been diminished by civil service reform. Of the legion of Irish-run machines, only Chicago and Albany remain as relics of the past. In all likelihood, these two vestiges will soon pass from the scene. The powerful Chicago machine has been progressively weakened since Richard Daley's death in 1976, losing the mayoral elections of 1979 and 1983. The Albany machine entered an interregnum phase with O'Connell successor Erastus Corning's death in 1983.

Paradoxically, the Irish machine's demise has been accompanied by a metamorphosis in our understanding of its achievements. During its heyday, it was castigated by Progressives as corrupt and undemocratic. In its twilight era, social scientists such as Robert Merton and Robert Dahl advanced a much different understanding of its performance. In the new rainbow theory, machines economically and politically incorporated working-class immigrant groups. Tammany Hall and the other big-city machines supposedly mobilized the Irish in the late nineteenth century, rewarding them with patronage jobs. In this fashion, machines served as a major route of Irish economic mobility. The Irish-run machines, in turn, supposedly fostered the political assimilation of the later-arriving Southern and Eastern Europeans by naturalizing and registering the new arrivals and by grooming their aspiring leaders for important party and governmental positions (Merton, 1968: 125-136; Dahl, 1961: 36-51).

The new interpretation exaggerates the ethnic benefits of machine politics. The early machines offered limited opportunities for Irish *group* social mobility. Nineteenth-century machines could not serve as channels of group economic advancement because of the political constraints upon their supply of resources. Yankee Republicans had not yet migrated in large numbers to the suburbs; they vigorously contested city elections. Machine bosses also had to contend with opponents in their own ranks: Democratic businessmen reformers advocating "tight-fisted" fiscal policies. This bipartisan conservative coalition forced the nascent Irish machines to pursue cautious taxation, spending, and indebtedness policies. Republicans dominated state politics during this era, reinforcing machine fiscal conservatism. Republican governors and legislators fashioned constitutional straitjackets to limit severely the machine's ability to raise taxes or increase municipal debts (Yearley, 1970: 3-74; Erie, 1983: 29-39).

Electoral limits on increasing tax rates and spending coupled with legal constraints upon city taxation and borrowing authority limited the nineteenth century machine's patronage supply. Only a few Irish could be given public jobs. Between 1870 and 1900 first- and second-generation Irish were rewarded with over one-third of the 150,000 newly created public jobs in the cities over 100,000 population. Yet in 1900, because of the limited scale of the public sector, only 6 percent of the urban Irish directly worked for government (Erie, 1978: 280-284).

Despite the building of the machines and the group allocation of available resources, the Irish community in the late nineteenth century was plagued by political tensions, especially between party and trade union leaders and between machine politicians and the Irish electorate. The fragile machine's maintenance needs—building citywide electoral majorities, placating tax-conscious businessmen, securing necessary party financing—had introduced a conservative strain into early Irish urban leadership. As they learned to manipulate the levers of urban power, Irish bosses turned their backs on more radical working-class ventures.

The first generation of Irish machine leaders risked working-class and immigrant political insurgency because of their conservative policies. Working-class people formed the largest component of the Irish-American community. Their political support was essential to the Democratic machines. Yet the Irish working class revolted against conservative Irish bosses. In New York City, Tammany boss John Kelly reduced city employment and wages during the depression-ridden 1870s. As a result, John Morrissey and the "short-hair" or lower-class Irish faction in Tammany broke with Kelly, claiming the boss was no longer sensitive to the needs of working-class Irish. Tammany lost the 1875 mayoral election in part because of erosion in its working-class and immigrant support. In the 1886 New York mayoral election, many working-class Irish deserted Tammany and supported Henry George, the United Labor Party candidate, nearly toppling Tammany. In San Francisco, Irish boss Chris Buckley pursued conservative fiscal policies and suffered a critical loss of Irish working-class support in the late

1880s (Genen, 1971: 92, 136; Shefter, 1978: 290; Bullough, 1979: 97, 160-163, 232-233).

Twentieth-century machines did a better job in economically aiding the Irish. Political and legal constraints upon the machine's ability to raise and spend money—and thus to create public jobs—began to ease as the machine's Republican and reform opponents moved to the suburbs, as state fiscal restraints were reduced with the advent of home rule, and as new services were needed by the millions of Southern and Eastern Europeans filling the cities. While machines directly or indirectly controlled less than 10 percent of urban job growth between 1870 and 1900, the newer machines directly or indirectly controlled the allocation of over one-quarter of big-city job growth between 1900 and 1930. In Irish-controlled machine cities such as New York, Jersey City, and Albany, first- and second-generation Irish were rewarded with between 50 and 60 percent of the urban public jobs created between 1900 and 1930. As a result, on the eve of the Depression, between 33 and 40 percent of the Irish stock work force in machine cities directly or indirectly depended upon machine-controlled patronage jobs (Clark, 1975: 341-342; Erie, 1984a; 12-15, 21-22).

With this sizable patronage system, the twentieth-century machines cemented the loyalties of the Irish working class and defused the threat of insurgency. In New York, the last serious threat of Irish insurgency against Tammany occurred in the 1905 mayoral election. Portraying himself as an urban populist, William Randolph Hearst ran against Tammany and almost won, running even with Tammany candidate George McClellan in working-class Irish districts. After the election, Tammany created thousands of municipal sinecures for the Irish by converting private utilities to public ownership. Thereafter, Irish discontent with Tammany would come from the reform-minded middle class rather than the working class (Connable and Silberfarb, 1967: 243; Myers, 1971: 309).

The twentieth-century machine's patronage policies appear to support the theory that politics served as an important conduit of Irish economic advancement. Yet compared with the Jews, the Irish were slow to build a middle class anchored in business and the professions. The Irish middle class was only emerging on the eve of the Depression; its arrival would not occur until after World War II. In light of their political success, why was middle-class status so slow coming for the Irish? Was there an inverse relationship between Irish political success and economic advancement? The Irish crowded into the largely blue-collar urban public sector in the late nineteenth and early twentieth centuries. Yet as low-paid policemen, firemen, and municipal clerks, the Irish were solidly lower-middle rather than middle class. It may be argued that by channeling so much of their economic energies into the public sector, the Irish forsook opportunities in the private sector save for industries such as construction that depended upon political connections. As Daniel Patrick Moynihan has accurately observed, the economic rewards of America have gone to entrepreneurs, not to functionaries. Moreover, Irish public sector job gains were fragile. The Depression forced cities to cut their payrolls. The 1930s also witnessed the long-awaited revolt of the Southern and Eastern Europeans against their Irish political overlords. Thousands of Irish-American payrollers lost their jobs due to retrenchment or machine overthrow, further delaying the building of an Irish middle class. With lessened job dependence upon the machine in the post-World War II era, however, the Irish rapidly moved into business and the professions (Glazer and Moynihan, 1964: 229, 259-260; Greeley, 1972: 122-128; Greeley, 1976: 54-55; Ryan, 1983: 106, 149).

If recent scholarship has exaggerated the machine's economic benefits for groups such as the Irish, can the same be said for its political benefits? It is true that the fledgling machines of the nineteenth century massively mobilized the Irish. The famine Irish arrived as the parties were entering their modern or mobilization phase. In the 1840s and 1850s Irish immigrants served as electoral cannon fodder in the competitive bidding wars between Democrats, Whigs, and Republicans. As Irish allegiance to the Democratic party solidified, the embryonic machines actively worked to naturalize and enroll Irish voters. Tammany Hall opened its famous Naturalization Bureau in 1840. By 1844 the department had naturalized 8000 Irish-born. The tempo of machine-sponsored mobilization quickened in the 1850s and 1860s. Tammany naturalized an average 5000 Irish-born per year between the mid-1850s and mid-1860s. In 1868 Tammany naturalized 41,000 foreign-born—over one-half of them Irish—in order to win the gubernatorial contest and free the city from Republican state control. Party-sponsored mobilization resulted in high electoral participation rates for the Irish. In San Francisco, for example, the proportion of Irish-born adult males registered to vote in the late nineteenth century nearly doubled that of the city's other foreign-born adult males—70 percent versus 37 percent—and equaled the rate for the native population (Connable and Silberfarb, 1967: 217; Kelley, 1974: 154-155, 172, 187; Erie, 1978).

Yet the early twentieth-century Irish machines did little to mobilize the Southern and Eastern European immigrants. In Irish-controlled cities such as New York, Boston, and Jersey City, naturalization and registration rates for the newcomers were quite low before the late 1920s. In New York City, for example, only 37 percent of the Russian Jews and 25 percent of the Italians were naturalized by 1920. Electoral participation rates for the second-wave migrants only increased in the late 1920s and 1930s, a response more to national candidates and issues than to the activities of local machines (Erie, 1984a: 25-26).

With so much of Irish economic well-being and group identity dependent upon continued control of the machines, the Irish were understandably loath to share power with the newcomers. Irish rule in New York, Boston, or Jersey City resembled British colonial rule in India. In order to preserve their hegemony, the Irish consciously pursued a divide and conquer strategy, pitting one ethnic leader or group against another. As the Southern and Eastern Europeans slowly mobilized, the Irish accommodated them in a parsimonious fashion, dispensing symbolic recognition and social services rather than the more valuable benefits of patronage and power. At critical moments, the Irish formed strategic alliances with some groups and not with others. As Jews and Italians flexed their political muscle in New York, the Irish were forced to choose which group would become a junior coalition partner, and thus eligible for a greater share of patronage and power. In New York, Jews became Tammany's chosen people in the 1920s. While the Irish offered the Jews a greater share of municipal employment, particularly in the rapidly expanding school system, Irish bosses worked as actively to reduce Italian influence by adroitly gerrymandering Italian neighborhoods (Lowi, 1964; Adler, 1971: 20-23, 48, 88-100, 197-199, 277; Henderson, 1976: 4-5, 137).

With the national mobilization of the new ethnics in the late 1920s and 1930s, the entrenched Irish machines were vulnerable to overthrow. In 1933 Tammany fell to reformer Fiorello La Guardia and his coalition of Italians and Jews. In 1949 the Hague machine in Jersey City fell to Italian and Polish insurgents. Yet in other cities the national political and economic upheavals of the 1930s presented machine-building opportunities for a new generation of Irish party bosses. Entrenched Republican machines in cities such as Chicago, Pittsburgh, and Philadelphia had been weakened by the Depression and national party realignment. As a minority party in these cities in the 1920s, the Democrats had been forced to more actively court and reward the new ethnics. Appealing to the second-wave immigrants, a new generation of

Irish bosses such as Ed Kelly, Pat Nash, and Richard Daley in Chicago and David Lawrence in Pittsburgh constructed more broad-based Democratic machines.

Postwar Irish machines such as the Daley organization in Chicago accommodated Southern and Eastern Europeans in different and less costly ways than the prewar machines had accommodated the Irish. The prewar high-tax job-creating policies had attracted considerable working-class Irish support. Wartime and postwar prosperity benefited the new ethnics, propelling large numbers into the property-owning middle class with less interest in public jobs. As homeowners, White ethnics objected to high property taxes. The Southern and Eastern Europeans demanded a new set of machine policies: low property taxes, the preservation of property values and White neighborhoods, and homeowner rather than welfare services, that is, clean streets, tree trimming, and garbage collection.

Postwar Irish-led machines accommodated the new ethnics' taxation and service demands—as long as Irish control over patronage and power was not seriously disturbed. In Chicago, Daley froze the local property tax rate in 1970 to placate White ethnic homeowners. Precinct workers were instructed to expedite homeowner services for Democratic voters. Yet the Irish retained control of key party positions and those city offices with major policymaking or patronage-dispensing responsibilities. The Windy City's Irish held the mayor's office from 1933 to 1983 (with a three-year hiatus) and retained control of the party organization until the early 1980s (Rakove, 1975: 34-41; Whitehead, 1977: 355-356).

Contrary to the rainbow theory, there were limits to the machine's economic and political incorporation of the European immigrant groups. The nascent nineteenth-century machines politically incorporated the Irish. Yet the economic disadvantages suffered by the Irish could not readily be overcome by politics; they may even have been aggravated. Most of the Irish seeking their fortune in the public sector walked into a blue-collar cul-de-sac. The early twentieth-century Irish machines did little to mobilize and reward the new arrivals from Southern and Eastern Europe. Even the postwar machines' apparent accommodation of the second-wave migrants was limited: low taxes and services rather than patronage and power.

Machines conservatively represented working-class as well as ethnic interests. Machine policies segmented the urban working class into incorporated and nonincorporated sectors—for instance, the Irish versus Jews, Italians, and Slavs—and encouraged demands for divisible rather than collective benefits, such as patronage rather than welfare programs. It is true that the patronage-based machines also supported Progressive and New Deal social welfare legislation: workman's compensation, factory safety legislation, the Wagner Act, and federal emergency jobs programs such as the WPA. Yet the Irish machines supported these social programs in order to appeal to the Southern and Eastern Europeans at minimal cost to Irish control over the machine's core resources of patronage and power. The machine pattern of selective ethnic mobilization and tangible reward continued with the later-arriving Blacks but with one important difference. In cities such as Chicago, the machine mobilized Blacks only when threatened with White ethnic insurgency. As the machine secured its White ethnic base with low taxes and homeowner services, it dispensed with Black political participation. As Blacks grew militant in the 1960s, the aging Irish machines piggybacked welfare-state programs such as Aid to Families with Dependent Children, Public Housing, and Community Action in order to reward—and control—Blacks at minimal cost to the city treasury and to White ethnic taxpayers.

What lessons can the Irish machines furnish for the recent Hispanic and Asian migrants to the cities? Before working-class groups such as Mexican Americans attempt to build their own machines, they would do well to study closely the two faces of machine power: They are institutions of economic reward but also of social control. Machines rewarded the Irish, but failed to mobilize and extend tangible benefits to other working-class ethnic groups. Moreover, as Ira Katznelson (1976: 224-225) has argued, machines inhibited other forms of working-class political action such as labor parties and unions. Irish machines were antagonistic toward labor parties and sometimes even toward the fledgling trade union movement (notwithstanding its Irish Catholic flavor in the big cities). Irish-controlled police departments were frequently used to break up socialist meetings, to scatter pickets, or to escort scabs. Irish machine antipathy was more political than cultural, a jurisdictional battle for working-class loyalties rather than a Catholic predilection for social stability and order (Erie, 1980: 270-271).

JEWS AND ITALIANS:
DELAYED MOBILIZATION AND ANTI-IRISH INSURGENCY

Fleeing economic and political conditions in Europe, a second wave of Southern and Eastern European migrants began arriving in large numbers in the late 1880s. Jews and Italians predominated among the new immigrants. By 1920, some 1.4 million Russians (roughly three-quarters Jewish) and 1.6 million Italians had migrated to the United States. Like the Irish before them, the newcomers congregated in the big cities. By 1920, 94 percent of the Russian Jewish migrants and 57 percent of the Italian-born lived in the nation's 25 largest cities.

Much like Mexican Americans today, the second-wave migrants politically mobilized slowly, particularly the Italians and Poles. Naturalization rates remained low until the late 1920s. The newcomers had low voter registration rates as well. In San Francisco, for example, only 39 percent of the naturalized Russian Jews were registered to vote in the early twentieth century compared with nearly three-quarters of the Bay Area's naturalized Irish.

Benign neglect on the part of the Irish machines only partially accounted for these low voter participation rates. The newcomers also mobilized more slowly than the Irish because of attributes brought from the Old World. Eastern European Jews, for example, came with none of the political advantages the Irish brought. Few spoke or read English. Most arrived with parochial identities formed in the *shtetl* or village. There were also strains between the new arrivals from Russia and Poland and the older, more acculturated German Jews. The forging of a common Jewish political consciousness would occur *after* migration.

The obstacles to Italian political mobilization were even greater than those facing the Jews. Like the Jews, migrating Italians lacked familiarity with the language and culture. Like the Jews, Italians were immersed in a web of local and family rather than national identities. In Southern Italy and Sicily, geographical and language barriers promoted a web of local loyalties and identities known as *campanilismo*. Unlike the Jews, but like many Hispanics today, large numbers of Italians migrated with the intent of ultimately returning to their homeland. Fleeing the pogroms carried out after the assassination of Czar Alexander in 1881, Jews sought permanent refuge, bringing their families with them. Italians, however, primarily migrated for economic rather than political reasons. Working-age males without families constituted 80 percent of the Italian migration prior to 1914. The scale of return migration to Italy seriously retarded the building of group political influence. In all, 42 percent of the Italian immigrants arriving before World War I returned to Italy.

Tightened naturalization and suffrage laws created additional barriers to electoral participation by the Southern and Eastern Europeans. The federal Naturalization Law of 1906, tightening and lengthening the naturalization process, was passed in the wake of growing resentment against the new immigrants. Radicals were disqualified. An English-speaking requirement was introduced. The new law required more stringent proof of lawful entry into the country and of five years continuous residence. Under the new law, the denial rate on naturalization petitions tripled to 14 percent while the average naturalization interval lengthened from five to eleven years (Gavit, 1971: 77-142, 232-241).

Republican-dominated state legislatures in the Northeast and Midwest were simultaneously tightening state voter qualification laws in order to make it more difficult for the Democratic machines to mobilize the new immigrants. The state suffrage restriction movement took a variety of forms: the abolition of alien suffrage, the imposition of literacy tests, and personal-registration requirements applied only to urban voters.

With these new formal barriers to electoral participation and with the Irish firmly controlling the urban Democratic party and unwilling to share power with the newcomers, the second-wave migrants turned inward to forge common political identities. In New York, Jewish political energies were channelled into the Socialist Party and into the unionization of the garment industry, which employed more than one-half of the city's Yiddish-speaking immigrants by 1910. Leadership of these early unions was heavily Socialist. Socialist Labor Party leaders such as Morris Hillquit formed the first Jewish union, the United Hebrew Trades, in 1888. A coordinating organization, the UHT quickly formed 32 unions (Howe, 1976: 80-84, 101-115, 287-324).

Jewish Socialists organized the needle trades as the Irish leadership of the skilled trades largely ignored unskilled and semiskilled workers. In New York City between 1909 and 1913, union membership in the needle trades grew from 30,000 to 250,000. The United Garment Workers, International Ladies Garment Workers, and the Amalgamated Clothing Workers won higher wages, shorter hours, and improved working conditions. Increasingly, Jewish garment workers thought of themselves politically in terms of occupation and class, signaling a further estrangement from the ethnically conscious machines (Kessner, 1977: 28-31).

Jewish political energies were also directed at creating the *landsmann* or mutual aid societies, which provided medical insurance, death benefits, and social activities for their members. These organizations were provincial, further isolating participants from the urban party system. *Landsmann* members came from the same town. Even the subsequent development of national *landsmannshaften* federations reinforced the isolation of Jews from the urban political system.

When Jews finally entered the larger urban political arena, they initially resisted the limited political blandishments of Irish machine politicians. Writing in the early twentieth century, Robert A. Woods (1903: 64), a Boston social worker, observed that "the Jew is a thorn in the flesh of the Irish politician." Jews were the chief supporters of New York City's powerful Socialist Party. In 1914, for example, the predominantly Jewish lower East Side sent a Socialist to Congress and several Socialists to the State Assembly. Jews were also active supporters of Progressive candidates. In New York, reformers Theodore Roosevelt and Charles Evans Hughes received large numbers of Jewish votes (Fuchs, 1950: 56-63, 123, 135-139).

Jewish support for radical and reform candidates forced the Irish machines leftward. As previously mentioned, machine support of labor and social welfare legislation during the Progressive Era can partially be viewed as a political strategy designed to undercut the appeal of Socialists and Progressives to the new immigrants. In New York, Tammany introduced and secured passage of important labor measures. Tammany's representatives dominated the state's Factory Investigating Commission, established following the disastrous Triangle Shirtwaist Company fire in 1911. The Commission proposed adding over fifty new laws to the state's labor code. The Tammany-controlled state legislature enacted nearly all of the commission's proposals. Tammany also supported social welfare legislation in order to appeal to the new immigrants. Tammany's Al Smith, the Assembly majority leader, and Robert Wagner, the Senate majority leader, secured passage of a widow's pension measure, workman's compensation, insurance and utility regulation, stricter tenement laws, and the establishment of a state employment agency (Huthmacher, 1962: 234-238; Weiss, 1968: 81-85).

Machine support of social welfare legislation was partially responsible for a temporary cessation of hostilities with the new immigrants. Jews and to a lesser degree Italians would be drawn into the machine fold during the early 1920s. Tammany's Al Smith ran for governor in 1919 and 1924, and cut heavily into the Jewish Socialist vote in New York City (Fuchs, 1950: 66; Henderson, 1976: 293-295).

The Irish machine's newfound Progressivism delayed but did not prevent insurgency by the new immigrants. The failure of the Irish to share machine power and patronage with the newcomers influenced the form ethnic insurgency would take. An anti-Irish revolt would take the form of an independent reform movement against machine politics.

The real reform challenge of the new immigrants to the Irish machines would occur in the 1930s and 1940s in response to national as well as local forces. The Depression and New Deal party realignment threatened the established Irish machines by upsetting the equilibrium between the party's management of electoral and patronage-distributing processes. The Depression depleted the machine's patronage stock while the national party realignment increased the number of new ethnic voters. In the 1928 presidential election, Democrat Al Smith's candidacy mobilized nonvoting Italians, Poles, Jews, and other big-city immigrants. The urban turnout surge was remarkable. In New York City, the number of voters increased by 38 percent between 1924 and 1928—from 1.4 million to 1.9 million—after having increased by only 15 percent between 1920 and 1924. Comparable presidential turnout surges between 1924 and 1928 occurred in other machine cities such as Boston and Chicago. Franklin Delano Roosevelt would build in 1932 and 1936 upon the big-city ethnic coalition fashioned initially by Al Smith. In New York City, the number of presidential voters rose by another 26 percent between 1932 and 1936; in Boston, by 22 percent. In all, the national political realignment of 1928 to 1936 contributed to an effective doubling of the number of voters in cities like New York and Chicago. The national mobilization of the new ethnics presented a potential electoral challenge to the entrenched Democratic machines to the extent the presidential turnout surge spilled over into local contests and was captured by reformers (Lubell, 1965; 43-55; Andersen, 1979: 19-52; Erie, 1984b: 11-12).

Reform leaders in such cities as New York and Jersey City capitalized upon the long-standing grievances of the Southern and Eastern Europeans with the Irish machines. In New York City, Fusionist candidate Fiorello La Guardia's strategy in the 1933 mayoral election was to mobilize Jews and Italians. Jews constituted over one-quarter and Italians one-sixth of the city's eligible electorate. La Guardia won in 1933, relegating Tammany to twelve years in the political wilderness, by mobilizing the new ethnics and by capitalizing upon a split in the traditional machine vote (Mann, 1965: 26, 113, 124, 138-146).

La Guardia's first administration was directed at solidifying new immigrant support to prevent Tammany's reemergence. La Guardia immediately embarked on a campaign to increase the number of Jews and other new ethnics on the city's payroll. La Guardia reformed the antiquated civil service system in order to end job discrimination against the non-Irish. Under the "Little Flower," the proportion of city positions covered by civil service rose from 55 percent in 1933 to 74 percent in 1939. The reform mayor also expanded the size of the city's human services bureaucracies (and thus the number of job opportunities for the new ethnic claimants). Propelled by increased funding as well as by population growth, the city payroll increased by 60 percent—from 86,000 to 137,000—during La Guardia's three terms (Garrett, 1961: 134, 147, 366; Bayor, 1978: 127, 135).

By the tandem tactics of civil service reform and bureaucratic expansion, reformer La Guardia was able to alter dramatically the ethnic composition of the city's bureaucracy, particularly in the "new" social service agencies rather than in the "old" police and fire departments where the Irish remained firmly entrenched. The Jews were the prime beneficiaries of La Guardia's personnel and social policies. In the city's school system, which the Irish had dominated until World War I, 56 percent of the entering teachers in 1940 were Jewish. La Guardia's policies produced the desired electoral reults. A reform coalition of the Southern and Eastern Europeans had been fashioned that would rule New York until the mid-1940s. Using many of La Guardia's campaign tactics, a similar reform coalition of the new ethnics overthrew the Hague machine in Jersey City in 1949 (Bayor, 1978: 26; Connors, 1971: 142-183; Kincaid, 1981: 502-511).

In cities such as Chicago, which did not have entrenched Irish machines, the Southern and Eastern Europeans were drawn into machine rather than reform politics in the 1930s. During the 1920s, under intense competitive pressures from Republican Mayor "Big Bill" Thompson, the Irish-controlled Chicago Democratic party significantly expanded its ethnic base. Between 1922 and 1930, for example, the non-Irish proportion of the party's central committee rose from 39 to 60 percent. The greater sharing of power and patronage with the new immigrants did not prevent a challenge to Irish power. Because the new ethnic party leaders had been given the requisite power and patronage with which to mount an intraparty challenge, the new immigrants' insurgency was channeled *within* the embryonic Democratic machine rather than into reform politics (Gottfried, 1962: 172-175, 326; Gosnell, 1968: 13, 109, 173; Allswang, 1971: 40-47, 105-106).

Yet the integration of the Southern and Eastern European immigrants into the Irish-run machines built during the New Deal era remained incomplete. In the 1950s and early 1960s the Chicago machine again faced White ethnic insurgency. Angered by a 100 percent increase in property taxes between 1955 and 1963, White homeowners, particularly Poles, revolted against Mayor Richard J. Daley, voting for insurgent Benjamin Adamowski in the 1963 mayoral election. Daley narrowly defeated Adamowski only because of a massive Black vote in machine-controlled wards. Thereafter, Daley pursued low-tax and homeowner-service policies in order to soothe Irish-Slavic tensions (Royko, 1971: 129-132; O'Connor, 1975: 149, 170-176).

In the Eastern cities where they had joined together to support successful reform movements in the 1930s and 1940s, Italians and Jews took divergent political paths in the postwar era. The Italians took over the weakened Irish machines while Jews remained in the reform camp. In New York City, a decaying Tammany remained out of power from 1933 to 1945. Tammany's aging Irish chieftains were forced to share power with the Italians in exchange for desperately needed funding. As Tammany turned for revenue to the Italian-controlled underworld, the organization's ethnic balance of power dramatically shifted. Gangsters such as "Lucky" Luciano and Frank Costello installed their own Italian district leaders. Led by Carmine De Sapio, an Italian Tammany bloc successfully challenged Irish party hegemony in the late 1940s (Bell, 1953: 131-154; Moscow, 1971: 54-59).

De Sapio employed both patronage and policy to rebuild an Italian-controlled Tammany. Tammany's mayoral candidate Robert F. Wagner, Jr., won in 1953. Under Wagner, municipal employment rose by nearly one-third to 262,000 in 1957. Tammany also embraced such popular reforms as public housing, rent control, civil rights and permanent voter registration (Sayre and Kaufman, 1960: 48; Moscow, 1971: 105, 114-115, 127).

Yet Italian machine-rebuilding efforts failed in New York. Despite Tammany's metamorphosis from patronage to policy, the city's Jews remained in the ranks of the reformers, organizing anti-Tammany political clubs throughout Manhattan in order to infiltrate and capture the party bureaucracy. Loss of state and federal patronage in the late 1950s and early 1960s seriously weakened Tammany's ability to fight the reformers. As Tammany weakened, Mayor Wagner severed his connection with De Sapio and joined the reformers. In the crucial 1961 mayoral election, the reformers defeated Tammany by capturing the votes of Jews, Blacks, and Puerto Ricans. An Italian-run Tammany would never again be resurrected (Wilson, 1966: 32-64; Moscow, 1971: 159, 166-172).

The urban political experience of the second-wave of European immigrants thus can be divided into three stages: a lengthy initial stage of nonincorporation analogous to the situation of Mexican Americans in cities such as Los Angeles today; a second stage of massive mobilization and insurgency channeled both within and against the Irish machines; and a third stage of ethnically differentiated incorporation into machine (Italians) and reform (Jewish) politics.

The Southern and Eastern European experience can yield important lessons for fourth-wave migrants, particularly for Mexican Americans. As did the second wave of migrants, Mexican Americans have found their path to urban power blocked by naturalization and electoral barriers. Mexican American political influence has been further diluted by high return migration rates to Mexico, as Italian American influence was similarly lessened in the early twentieth century. Thus the slow mobilization of Mexican Americans today closely resembles the delayed incorporation of groups such as the Italians and Poles. With the Irish blocking their way, the Southern and Eastern Europeans required forty years after their arrival in order to wrest control of the levers of urban power. Like Mexican Americans, the second-wave migrants turned to institution-building within their communities before gaining influence in the larger urban arena. For the Southern and Eastern Europeans, the ascension to urban power was not an incremental process. Ethnic revolt occurred suddenly and unexpectedly, a product of determined ethnic political leadership and of national political and economic forces promoting group electoral participation. Massive group mobilization represented the key ingredient in winning the early ethnic wars of urban succession.

Ironically, in the process of wresting power from the Irish, the Southern and Eastern Europeans created additional barriers for later-arriving Blacks, Hispanics, and the other newer immigrants. In their battle against the Irish machines, the second-wave ethnics joined Yankee reformers in bringing to the Eastern cities the reforms fashioned by the Progressives in Western cities such as Los Angeles: nonpartisanship, at-large councilmanic elections, expanded civil services coverage and educational requirements for city employment. Designed to prevent the reemergence of machines, reforms also consolidated the power of the new ethnics by making it more difficult for working-class Blacks and Hispanics to enter the urban political arena.

BLACKS AND WELFARE-STATE PROGRAMS

Between 1940 and 1980 some 5 million Southern Blacks migrated to the Northern cities. Agricultural modernization in the South coupled with changes in the Northern urban economy both contributed to the massive northward exodus. In the rural South, mechanization encouraged large-scale capital-intensive farming, disrupting Black tenant farming arrangements. In the urban North, wartime labor shortages created job opportunities for displaced Blacks. In the postwar era, as manufacturing and as newer industries such as aerospace and electronics both followed suburban locational patterns, other sectors of the urban economy, for instance, government and services, created additional jobs for Black workers. The Black migration dramatically changed the complexion of the Northern cities. In New York City alone, the Black population increased by over 1 million—to 1.8 million—between 1940 and 1980. By 1980, Blacks made up over 70 percent of Detroit's population and 40 percent of Chicago's and Philadelphia's (Mollenkopf, 1983: 20-28).

In the pre-1960 period, Black submachines to the aging White machines had emerged in cities such as Chicago. Congressman William Dawson, the only Black in the Daley machine's inner circle, ran the Black submachine in the South Side ghetto. Chicago's Blacks supplied the margin of victory in three of Mayor Daley's six elections. Yet Blacks received few direct material rewards from the machine. Rather than giving Dawson more power or patronage, Daley rewarded him with control of the numbers racket. By the late 1970s, Blacks made up 40 percent of the Windy City's population, yet represented only 20 percent of the municipal work force, largely in menial positions (Gosnell, 1935; Rakove, 1975: 16, 110-111; O'Connor, 1975: 120-121; Kilson, 1971: 182-189).

Black demands on the urban machines escalated during the turbulent 1960s. In Chicago, Black demands started growing in the early 1960s, led by Saul Alinsky's Woodlawn organization seeking better schools. In 1966, Martin Luther King, Jr., took the civil rights movement north to Chicago, demanding an end to segregated housing and discrimination in employment. Racial tensions increased as King marched through Chicago's White suburbs. Riots rocked the Windy City in 1966 in the Lawndale ghetto and in 1968 in the West Side ghetto.

Machines such as the Daley organization judiciously used welfare-state programs to control the minority vote and to siphon off discontent at minimal cost to the city treasury and to tax-conscious White homeowners. Both New Deal and Great Society social programs were used for these purposes. Public Housing and Aid to Families with Dependent Children (AFDC) represented the two major New Deal programs machines used to placate Blacks.

Migrating Blacks confronted an acute housing shortage as Northern machines collaborated with real estate brokers to confine Blacks to inner-city ghettos. By reinforcing existing racially segregated residential patterns, machines built support in White ethnic neighborhoods. Machines also collaborated with downtown businesses in pioneering urban renewal programs to revitalize the decaying central city. These programs contributed substantially to the inner-city housing shortage by uprooting thousands of Blacks. As the Northern ghettos filled in the postwar era, the big-city machines lobbied for federal housing money to build low-income public housing projects. Machines such as the Daley organization deliberately pursued a policy of racial containment, concentrating public housing projects in the ghetto. Segregated public housing projects not only soothed the fears of White ethnics; they also served to concentrate the Black vote and make it more controllable (Royko, 1971: 136-144; Hirsch, 1983).

The postwar suburbanization of private sector jobs and industry contributed to chronically high unemployment rates for urban Blacks. Rather than grant a greater share of municipal employment to the increasingly militant Black community, machines chose welfare as a politically cheaper response. As Piven and Cloward (1971: 242-243) have argued, welfare limited the scope of urban racial conflict, for this strategy did not necessitate challenging White political prerogatives. Machines exerted little control over determining AFDC eligibility. Yet there were ways of politicizing this collective benefit program. So that it could claim credit for increasing welfare benefit levels, the Chicago machine instructed its state and congressional legislative delegations to vote for welfare liberalization measures. At the local level, Chicago's precinct captains set up welfare information bureaus in public housing projects to put prospective welfare clients in contact with the appropriate social agency. Even though the Windy City's machine politicians did not control eligibility, they would and did threaten welfare recipients with loss of public assistance should they vote for the machine's opponents. Under machine auspices, the welfare participation rate in Chicago for Black families rose from 15 percent in the mid-1960s to 32 percent in 1979 (Banfield, 1961: 74; Royko, 1971: 138).

Machines also relied upon Great Society programs to build political support in and stabilize the Black community, particularly among the fledgling middle class. While the federal antipoverty programs of the 1960s such as Community Action, Job Corps, and Model Cities ostensibly were targeted at the poor, they created considerable employment for human service providers. For example, the $2.7 billion annual funding in the early 1970s for Title I of the Elementary and Secondary Education Act created 275,000 full- and part-time teaching and administrative positions. Nationwide, Great Society programs generated 2 million new human services jobs, primarily in local government and in community-based organizations.

Machine politicians had strong political incentive to use federally funded social welfare employment to respond to growing Black demands. As with welfare, federally funded social employment limited the scope of interracial conflict. By channeling Blacks to jobs in expanding social service agencies, machine politicians minimized conflict with Whites in traditional city agencies such as police and fire departments. Social welfare employment also represented one of the most tangible and effective ways that machine leaders could respond to the threat of urban disorder. In a city such as Chicago, where the machine controlled local antipoverty funding, over 40 percent of the Black job gains in the 1960s occurred in social welfare agencies. Nationwide, by the late 1970s, over one-half of the "new" Black middle class of professionals and managers worked in human services (Greenstone and Peterson, 1973; Brown and Erie, 1981: 302-304).

Machine politicians carefully ensured that the party organization and not the Black community controlled local antipoverty agencies. In Chicago, Mayor Daley successfully resisted Black political demands for control of the community action program by demonstrating the machine's importance in Democratic presidential politics. In 1964, Daley delivered Chicago for Lyndon Johnson by a record 675,000 votes in order to highlight the president's dependence upon the machine. By this show of electoral clout, reminiscent of the Kelly/Nash machine's mobilization of Chicago voters in the mid-1930s to capture control of the local WPA program, Daley received $140 million in federal antipoverty funds (O'Connor, 1975: 107; Protess, 1974: 184-202).

Local politicians in reform cities also used welfare-state programs to secure the Black vote and to reduce the level of unrest. In New York City, for example, reform Mayor John Lindsay used federal social programs to build a new political coalition in the late 1960s. Lindsay had narrowly won in 1965 with a makeshift coalition of traditional Republicans, Liberals, and reform Democrats. Lacking a broad and durable constituency, Lindsay used federally funded jobs to expand his control over the city administration and broaden his base to include Blacks and Puerto Ricans. Blacks were well organized in New York, and they were able to gain an extraordinary level of participation in and influence over the local antipoverty program. In all, Blacks received over 60 percent of the public sector jobs created in the city during the 1960s (Adams, 1976: 37; Morris, 1981: 25-26, 59-66, 126).

The growing dependence of both the Black middle class and poor upon welfare-state programs created an important, yet largely invisible, arena for urban Black politics in the 1970s. The visible arena was local elected office, and here urban Blacks made impressive strides. As previously mentioned, Black officials have been elected in numerous cities. There was a paradoxical character, however, to the political gains in the 1970s. As Blacks moved from protest to politics, as record numbers of Black local officials were elected, Black voter participation rates declined sharply. In Chicago, for example, the turnout rate for Blacks declined from 60 to 37 percent between 1964 and 1976.

Welfare-state expansion from the mid-1960s to the late 1970s contributed to this erosion of the urban Black electoral base. Minority social service providers increasingly involved themselves in bureaucratic politics within the intergovernmental grant system rather than in mobilizing their clientele. The expansion of means-tested programs such as AFDC depoliticized welfare recipients by isolating them from the work experiences encouraging political participation. Social programs thus represented a potent tool for local politicians to coopt the black middle class and depoliticize the underclass (Cloward and Piven, 1974: 16-25; Hamilton, 1976: 239-255; Brown and Erie, 1981: 322-328).

Ronald Reagan's election in 1980 led to an unprecedented attack on the social programs integrating the Black middle class and poor. For fiscal year 1982, the Reagan administration reduced federal social outlays by $35 billion, primarily in means-tested programs serving the poor. Budget reductions in subsequent years have been directed at the social service programs providing employment for minorities— education, manpower and training, health and social services. In fiscal year 1983, for example, Great Society service programs were the target of over $11 billion in cutbacks (Aaron et al., 1982: 149-150; Brown, 1982).

The Reagan retrenchment policies have galvanized the Black community in such cities as Chicago, Philadelphia, Boston, and New Orleans, uniting the middle class and poor and contributing to the voter mobilization campaigns of 1982 and 1983. Social program cutbacks encouraged both service providers and welfare recipients to politically mobilize. Threatened with job loss, service providers have incentive to organize both themselves and their welfare clientele in order to restore social program funding. Threatened with benefit loss, welfare recipients rediscover the linkage between political action and benefit receipt.

Black insurgency against the Chicago machine in 1983 illustrates the process by which social program cutbacks have led to Black remobilization. The 1975 mayoral election revealed the first significant Black voter disenchantment with the Daley machine. The machine's personnel and police policies, not social program reductions, prompted these initial stirrings of minority discontent. Capitalizing upon Black service grievances, maverick Jane Byrne defeated machine candidate Michael Bilandic in the 1979 mayoral election with 63 percent of the Black vote (O'Connor, 1977: 50, 121; Preston, 1982: 168-174).

Despite growing discontent with the Chicago machine's policies, the Black vote was not fully mobilized until *after* Reagan's attack on the welfare state. Black voter participation rates were low prior to 1980. As of 1979, only 400,000 of the city's 950,000 eligible Black voters were registered. Reagan's policies, however, brought the Black middle class back into electoral politics. Blacks working in social service agencies, activists in community-based organizations hurt by funding cutbacks and ministers orchestrated a large-scale voter registration drive among the underclass in churches, welfare, and unemployment offices. Between 1981 and 1983, the number of registered Black voters in Chicago dramatically rose from 400,000 to 600,000 (Marable, 1983: 57-60; Whitehead, 1983: 30; Thompson, 1984: 4).

In the 1983 mayoral election, newly enrolled Blacks rallied behind Black candidate Harold Washington. In the Democratic primary, Washington defeated Mayor Byrne and State's Attorney Richard M. Daley, son of the late mayor. In the general election, Washington narrowly defeated the Republican candidate with a record 80 percent turnout of Black voters. In Philadelphia, a similar mobilization of Black voters in 1983 contributed to Black candidate Wilson Goode's mayoral victory over Frank Rizzo.

Thus the Black urban political experience differs considerably from that of the White Europeans. If machines directly shaped Irish-American politics and indirectly influenced the forms and timing of Southern and Eastern European insurgency, then Black politics has been shaped by the dynamics of the American welfare state. The cyclical character of urban black political participation—the erosion of a mass electoral base in the 1970s and its reemergence in the 1980s—reflects the larger contradictions of the welfare state. Social program growth has a tendency to be politically self-defeating because it erodes the political support of the groups it serves. Welfare-state contraction, however, reverses this cycle. Threatened with loss of jobs and benefits, providers and recipients have incentive to unite in collective political action.

The Black experience yields lessons for recent Hispanic and Asian urban migrants. Before fourth-wave migrants call for increases in federal funding for manpower and training, education, and social service programs, they would do well to examine closely the two faces of the welfare state. As with the machine, it is a mechanism of both social control and economic reward. Welfare-state programs economically rewarded the Black middle class at the cost of both creating and sustaining economic and political stratification *within* the Black community.

HISPANICS, ASIANS, AND URBAN POLITICS IN THE SUNBELT

Though large numbers of Hispanics have lived in the United States since the colonial period and Asians since the mid-nineteenth century, their numbers have increased substantially since the mid-1960s in response to political and economic developments in the Third World and to the liberalization of the nation's immigration laws. In 1965, U.S. immigration policy was tilted westward and southward—toward Asia and Latin America and away from Europe. Representing the most fundamental policy change since 1924, the new law resulted in a 50 percent increase in the number of legal immigrants. By the mid-1970s, 500,000 legal migrants were arriving annually, the largest number since the early 1920s. Nearly one-half of the post 1965 immigration has come from just five countries: Mexico, Cuba, the Philippines, Korea and Taiwan (Bryce-Laporte, 1982: 66-70). Millions of undocumented aliens also have joined the fourth-wave migration. The new migration promises to transform dramatically the political life of many of the nation's growing Sunbelt cities, particularly Miami, San Antonio, El Paso, and Los Angeles, where the new migrants approach 50 percent of the population (Crewdson, 1983: 279-285).

It is too soon to know what special forms urban political incorporation of the newly arrived Hispanics and Asians will take. Yet it is clear that the urban political experiences of the fourth-wave migrants will differ considerably from the Irish, Jews, Italians, and Blacks. The newest urban arrivals encounter quite different political structures than the party machine or welfare-state bureaucracy, regardless of their imperfections as instruments of ethnic incorporation. The old urban machines are dead and are unlikely to be revived. Ronald Reagan's continuing attack on welfare-state programs has drastically reduced federal spending for the poor. Special federal assistance programs for Cuban and Vietnamese refugees also largely have been terminated. Fourth-wave migrants thus come of age politically as the integrating institutions of the past—limited as they were—have been dismantled.

Moreover, the newcomers generally settle in Sunbelt rather than Frostbelt cities. The political structures of Sunbelt cities make ethnic political incorporation more difficult. Reformed governmental structures, for example, nonpartisanship and at-large elections, create obstacles to group mobilization. Sunbelt cities also have smaller public sectors and more stringent civil service systems, thus limiting government job opportunities for minorities.

There is evidence that reformed structures have delayed the incorporation of the fourth-wave migrants. In San Antonio, the Anglo-dominated Good Government League successfully implemented at-large councilmanic elections, assuring Anglo control of the city until the late 1970s. In nonpartisan Los Angeles, where Hispanics comprise nearly one-third of the population, only one Mexican American has served on the city council since the 1850s and none has ever served on the county board of supervisors (Johnson, 1983: 248-254; Garreau, 1981: 231-243; Clark, 1983: 300-303).

Reformed structures also have blunted Cuban influence in Miami. In 1957, Dade County residents adopted a two-tier system with a strong countywide metropolitan government, assuming expanded functions at the expense of the area's fragmented municipalities. Middle-class leadership and a large voting bloc made Cuban influence quickly felt in Miami *city* politics. Middle-class professionals and businessmen predominated among the initial waves of Cuban migration, from 1959 to 1962 and from 1965 to 1973. The transplanted Cuban middle class moved into the export trade business and helped transform Miami in the 1970s into an international trade and banking center for the Latin American market. By the early 1980s, the number of Cubans in Dade County had swelled to 600,000, representing over one-third of the county's and over one-half of the city's population. The Cuban vote, however, remained small during the 1960s. Cuban exiles anticipated a return to their homeland and thus did not seek U.S. citizenship. With the growing despair after the Bay of Pigs, however, thousands of exiles have applied for U.S. citizenship. By 1984, over 80 percent of Cuban-born adults were naturalized. In Miami city elections, the Hispanic political participation rate rose to twice that of Anglos. Cubans joined Anglos and Blacks to elect Puerto Rican Maurice Ferre mayor in 1973. By 1980, Mayor Ferre and two Cuban commissioners had placed

Miami's city government in Hispanic control. While Cubans have been successful in penetrating the city arena, they have been much less successful in penetrating the more powerful Anglo-controlled metropolitan arena (Llanes, 1982: 7-10, 98-100; Mohl, 1983: 70-76, 82-89; Crewdson, 1983: 284-285; Peirce and Hagstrom, 1983: 525-526).

Federal antidiscrimination policies have eliminated some of the structural barriers to political participation. The Voting Rights Act of 1965 abolished literacy tests, while the Twenty-Fourth Amendment curtailed poll taxes. Justice Department litigation has resulted in the substitution of district for at-large elections in several Southwestern cities and has prevented municipal annexations that dilute the Chicano vote (Moore and Pachon, 1985).

With the lowering of those barriers, fourth-wave political influence has grown in the Sunbelt cities. In San Antonio, for example, the Mexican American challenge to the Anglo power structure began in 1973 with the formation of Communities Organized for Public Services (COPS), an Alinsky-inspired church based organization focusing upon inner-city policy issues. Yet the city's Chicanos made little headway until 1977, when the city, under pressure from the Justice Department's Civil Rights Division, switched back to the district method for electing councilmen. Soon thereafter, five Mexican American councilmen had been elected. In 1981, San Antonio's voters elected Henry Cisneros as the first Mexican American mayor of a large U.S. city (Johnson, 1983: 248-254).

Demographic trends ensure that Hispanics will exercise a far greater degree of urban power in the future. The number of eligible Chicano voters will mushroom in the late 1980s and early 1990s. The Chicano community is young with an average age of 18 versus 30 for Anglos. In cities such as Los Angeles, over one-quarter of the Hispanic community lacks proper documentation, and thus is ineligible for citizenship and the vote; however, the sons and daughters of undocumented aliens born in the United States are citizens and thus ultimately will be eligible voters.

Given the concentration of Mexican Americans in the Southwest, and that region's growing importance in presidential politics, the national parties work to mobilize the Chicano urban vote. Of all Mexican Americans, 85 percent reside in the rapidly growing states of California and Texas. These two states alone have nearly one-quarter of the Electoral College votes needed for victory (Moore and Pachon, 1985). These calculations have not been lost on presidential candidates. In 1972, Richard Nixon pursued not only his famous Southern strategy but a Southwestern "Chicano" strategy as well. Nixon successfully wooed Chicano votes with high-level government appointments and targeted federal social spending. During the first Nixon administration, federal health and social expenditures targeted to Hispanic communities in Texas and California totaled $100 million. As a result, Nixon secured nearly one-third of the Hispanic vote (and roughly one-half in Texas) compared with the 10 to 15 percent normally received by Republican candidates. As competition for the Sunbelt intensifies, the parties will make major efforts to woo the Chicano vote, thus strengthening Mexican American big-city electoral clout (Castro, 1974: 198-214; Olson, 1979; Garcia and de la Garza: 1982: 94-102).

The Mexican American political experience is best understood when we realize that slow incorporation is the norm, not the exception, for working-class ethnic groups in the United States. The rapid mobilization and reward of the Irish is the real exception. Yet the supposed Irish political success story, no matter how costly in terms of the delayed incorporation of the Southern and Eastern Europeans or the diminished possibilities for more radical working-class politics, has become the yardstick with which to measure the political progress of all later-arriving groups. Small wonder that the legend of Irish power served as a crucial yardstick in the Black power debates of the 1960s because it supposedly demonstrated the efficacy of local electoral strategies for redistributing public sector resources, thus enabling significant numbers of an ethnic group to escape poverty (Cruse, 1967: 315-316; Stone, 1968: 110-118; Hamilton, 1974: 191-193). And now the slow incorporation of Chicanos is invidiously compared with that of Blacks notwithstanding the fact that the welfare-state structures incorporating urban Blacks may be politically self-defeating, and, in any event, are now being dismantled.

Unlike Hispanics, Asian-American political influence is being felt in suburban rather than urban politics. Since 1965 nearly 2 million Chinese, Japanese, Filipino, Vietnamese, and Korean migrants have settled in West Coast metropolitan areas. Over 700,000 Asian Americans live in Los Angeles County alone, representing 10 percent of the county's population. With sizable middle classes in small business and in technical fields, the Asian groups have moved to the suburbs, where they have become active in local politics. Middle-class Japanese Americans are influential in the Los Angeles suburb of Gardena, while immigrant Chinese and Japanese are involved in Monterey Park. Vietnamese refugees, the beneficiaries of a massive federal resettlement program, have clustered in the Orange County suburb of Westminster (Clark, 1983: 303-304; Olson, 1979; Montera, 1979).

In this essay I have compared the old and new urban ethnic politics, highlighting how political institutions—party machines, welfare-state bureaucracies, and reformed structures—shape ethnic behavior. Because of the failure to understand this institutional context adequately, past studies have misinterpreted the political experience of the earlier working-class ethnic arrivals. Limits on the machine's supply of resources forced the Irish to fight a rearguard action, hoarding limited gains against challenges by the new ethnic claimants. Thus, contrary to the rainbow theory, the old-style urban machines did not politically and economically incorporate the Southern and Eastern Europeans as they had the Irish. The economic integration of urban Blacks by the welfare state and its political consequences also have been little understood. As we better understand the institutional character of earlier ethnic incorporation and its costs as well as benefits, we better illuminate as well the political experiences of the more recent arrivals.

References

AARON, H. J. (1982) "Nondefense programs," in J. A. Pechman (ed.) Setting National Priorities: The 1983 Budget. Washington, DC: Brookings Institution.

ABAD, V., J. RAMOS, and E. BOYCE (1974) "A model for delivery of mental health services to Spanish-speaking minorities." American Journal of Orthopsychiatry 44 (July): 584-595.

ABBOTT, E. [ed.] (1926) Historical Aspects of the Immigration Problem: Select Documents. Chicago: University of Chicago Press.

ACUNA, R. (1972) Occupied America: The Chicano's Struggle Toward Liberation. San Francisco: Canfield.

ADAMS, J. R. (1976) "Why New York went broke." Commentary 61: 31-37.

ADLER, N. M. (1971) "Ethnics in politics: access to office in New York City." Ph.D. dissertation, University of Wisconsin.

AGUIRRE, A., Jr. (1982) "The political economy context of language in social service delivery for Hispanics," in W. A. Van Horne (ed.) Ethnicity and Public Policy. Madison: University of Wisconsin System.

AKULICZ DE SANTIAGO, A. (1984) Residential Segregation of Spanish Origin Populations: Recent Trends in a Sample of U.S. Cities. Ph.D. dissertation, University of Wisconsin—Milwaukee.

ALLSWANG, J. M. (1971) A House for All Peoples: Ethnic Politics in Chicago, 1890-1936. Lexington: University Press of Kentucky.

ANDERSON, K. (1979) The Creation of a Democratic Majority, 1928-1936. Chicago: University of Chicago Press.

ANDERSON, K. (1983) "The new Ellis Island." Time 121 (June 13): 18-25.

ANTUNES, G., C. GORDON, C. M. GAITZ, and J. SCOTT (1974) "Ethnicity, socioeconomic status, and the etiology of psychological distress." Sociology and Social Research 58: 361-368.

ARMINGTON, C. and M. ODLE (1982) "Small business—how many jobs?" Brookings Review (Winter): 14-17.

Aspira of America (1978) Social Factors in the Educational Attainment among Puerto Ricans in the U.S. Metropolitan Areas. New York: Author.

Aspira of New York (1983) Racial and Ethnic High School Dropout Rates in New York City: A Summary Report. New York: Author.

BABIGIAN, H. M. (1977) "The impact of community mental health centers on the utilization of services." Archives of General Psychiatry 34: 385-394.

BACH, R. L. and J. B. BACH (1980) "Employment patterns in southeast Asian refugees." Monthly Labor Review 103: 31-38.

BACHRACH, L. L. (1976) Deinstitutionalization: An Analytical Review and Sociological Perspective. Washington, DC: Government Printing Office.

BAILEY, T. (1984) A Case Study of Immigrants in the Restaurant Industry: Industry Structure and Labor Market Competition. New York: Conservation of Human Resources, Columbia University.

BAILYN, B. (1955) The New England Merchants in the Seventeenth Century. Cambridge, MA: Harvard University Press.

BALKAN, S., R. J. BERGER, and J. SCHMIDT (1980) Crime and Deviance in America. Belmont, CA: Wadsworth.

BANFIELD, E. C. (1974) The Unheavenly City Revisited. Boston: Little, Brown.

———(1961) Political Influence. New York: Free Press.

———and J. Q. WILSON (1963) City Politics. New York: Vintage.

BAYOR, R. H. (1978) Neighbors in Conflict: The Irish, Germans, Jews, and Italians of New York City, 1929-1944. Baltimore: John Hopkins University Press.

BECKER, H. J. (1979) Personal Networks of Opportunity in Obtaining Jobs: Racial Differences and Effects of Segregation. Baltimore: Center for Social Organization of Schools, Johns Hopkins University.

BELL, D. (1953) "Crime as an American way of life." Antioch Review 13: 131-154.

BELL, D. (1960) The End of Idology. New York: Free Press.

BELL, R., J. LEROY, E. LIN, and J. SCHWAB (1981) "Change and psychopathology: Epidemiological considerations." Community Mental Health Journal 17, 3: 203-213.

BEQUAI, A. (1979) Organizing Crime. Lexington, MA: D. C. Heath.

BERKMAN, P. L. (1971) "Measurement of mental health in a general population survey." American Journal of Epidemiology 94, 2: 105-111.

BILLINGTON, R. A. (1938) The Protestant Crusade, 1800-1860: A Study of the Origins of American Nativism. New York: Holt, Rinehart.

BLOOMBAUM, M., J. TAMAMOTO, and Q. EVANS (1968) "Cultural stereotyping among psychotherapists." Journal of Counseling and Clinical Psychology.

BLUESTONE, B. and B. HARRISON (1983) The Deindustrialization of America. New York: Basic Books.

BLUHM, D. A. (1983) "Stream of immigrants floods California services." Milwaukee Journal, September 12: 6.

BLUM, J. D., and F. C. REDLICH (1980) "Mental health practitioners: Old stereotypes and new realities." Archives of General Psychiatry 37: 1247-1253.

BOBCOCK, J. P. (1894) "The Irish conquest of our cities." Forum 17: 186-195.

BODNAR, J. E. (1976) "The impact of the 'new immigration' on the Black worker: Steelton, Pennsylvania, 1880-1920." Labor History 17(Spring): 214-229.

BONILLA, F. (1984) "Ethnic orbits: The circulation of capitals and peoples." Presented at the Conference on Ethnicity and Race in the Last Quarter of the Twentieth Century, State University of New York—Albany, April.

BORJAS, G. J. (1981) "Hispanic immigrants in the U.S. labor market: an empirical analysis," in M. Tienda (ed.) Hispanic Origin Workers in the U.S. Labor Market: Comparative Analyses of Employment Outcomes. Final report to the U.S. Department of Labor, Employment and Training Administration.

BORUS, J. F. (1981) "Sounding board: Deinstitutionalization of the chronically mentally ill." New England Journal of Medicine 305 (August 6): 339-342.

BOULDING, K. (1967) "The boundaries between social policy and economic policy." Social Work 12: 3-11.

BOUVIER, L. (1983) "U.S. immigration: effects on population growth and structure," in M. M. Kritz (ed.) U.S. Immigration and Refugee Policy: Global and Domestic Issues. Lexington, MA: D. C. Heath.

BOYD, M. (1974) "The changing nature of the central and southeast Asian immigration to the United States: 1961-1972" International Migration Review 8: 507-520.

———(1971) "Oriental immigration: the experience of the Chinese, Japanese, and Filipino population in the United States." International Migration Review 5: 48-60.

BOYER, E. L. (1984) "The test of growing student diversity." New York Times, education supplement (November 11).

———(1983) High School: A Report on Secondary Education in America. New York: Harper & Row.

BRAUN, P., G. KOCHANSKY, R. SHAPIRO, et al. (1981) "Overview: deinstitutionalization of psychiatric patients, a critical review of outcome studies." American Journal of Psychiatry 138: 736-749.

BREMMER, R. H. (1960) From the Depths: Discovery of Poverty in the United States. New York: New York University Press.

BRISCHETTO, R. et al. (1979) Minorities, the Poor and School Finance Reform: Report to the National Institute of Education. Washington, DC: Intercultural Development Research Association.

BRODY, D. (1969) Steelworkers in America. New York: Harper & Row.

BROOKS, G. (1984) "Offered sanctuary: scores of U.S. churches take in illegal aliens fleeing Latin America." Wall Street Journal (June 21).

BROWN, M. K. (1982) "Gutting the Great Society: Black economic progress and the budget cuts." Urban League Review 7: 11-24.

———and S. P. ERIE (1981) "Blacks and the legacy of the Great Society: the economic and political impact of federal social policy." Public Policy 29: 299-330.

BROWN, R. M. (1968) "Historical patterns of violence in America," in H. D. Graham and T. R. Gurr (eds.) Violence in America. New York: Signet.

BROWN, T. et al. (1973) "Mental illness and the role of mental health facilities in Chinatown," in S. Sue and W. Wagner (eds.) Asian Americans: Psychological Perspectives. Ben Lomond, CA: Science and Behavior Books.

BRYCE-LAPORTE, R. S. (1982) "The new immigration: its origin, visibility and implications for public policy," in W. Van Horne (ed.) Ethnicity and Public Policy. Madison: University of Wisconsin Press.

———(1979a) "New York City and the new Caribbean immigration: a contextual statement." International Migration Review 13, 2: 214-234.

———[ed.] (1979b) Sourcebook on the New Immigration to the United States. New Brunswick, NJ: Transaction.

BUCHANAN, S. H. (1979) "Language and identity: Haitians in New York City." International Migration Review 13, 2: 298-313.

BUI, D. (1983) "The Indochinese mutual assistance associations," in Special Services for Groups, Asian American Community Mental Health Training Center, Bridging Cultures: Southeast Asian Refugees in America.

BULLOGH, W. A. (1979) The Blind Boss and His City: Christopher Augustine Buckley and Nineteenth-Century San Francisco. Berkeley: University of California Press.

BURGESS, E. W. and D. J. BOGUE (1964) "Research in urban society: a long view," in E. W. Burgess and D. J. Bogue (eds.) Contributions to Urban Sociology. Chicago: University of Chicago Press.

California Department of Education (1982) California Public Schools: Enrollment by Ethnicity of the 25 Largest Districts in the 1981-82 OBEDS Data Base. Sacramento: Author.

————Bilingual Education Office (1984a) LEP Enrollment: District/County Summary. Sacramento: Author.

————(1984b) Districts Ranked by Enrollment of LEP Students. Sacramento: Author.

CARDOZA, D., L. HUDDY, and D. O. SEARS (1984) The Symbolic Attitudes Study: Public Attitudes Toward Bilingual Education. Technical Report R-21. Los Alamitos, CA: National Center for Bilingual Education.

CARHUFF, R. and R. PIERCE (1967) "Differential effects of therapist race and social class upon patient depth of self-exploration in the initial clinic interview." Journal of Consulting Psychology 31: 632-634.

CARR, L. and N. DRAUSE (1978) "Social status, psychiatric symptomatology, and response bias." Journal of Health and Social Behavior 19: 86-91.

CARTER, T. P. and R. D. SEGURA (1979) Mexican Americans in School: A Decade of Change. New York: College Entrance Examination Board.

CASTRO, T. (1974) Chicano Power: The Emergence of Mexican America. New York: Saturday Review Press.

CASUSO, J. and E. CAMACHO (1984) "Hispanics in Chicago: the Cubans." Chicago Reporter.

CHAMBLISS, W. (1978) "The political economy of smack: opiates, capitalism, and law." Research in Law and Sociology 1: 115-141.

CHASE, A. (1977) The Legacy of Malthus: The Social Costs of the New Scientific Racism. New York: Knopf.

CHASE, W. L. (1983) "Our big cities go ethnic." U.S. News & World Report 94 (March 21): 49-58.

CHISWICK, B. R. (1983) "An analysis of the earnings and employment of Asian-American men." Journal of Labor Economics 1: 197-214.

————(1980) "Immigrant earnings patterns by sex, race and ethnic groupings." Monthly Labor Review 103, 10: 22-25.

————(1979) "The economic progress of immigrants: some apparently universal patterns," in W. Fellner (ed.) Contemporary Economic Problems. Washington, DC: American Enterprise Institute.

————(1978) "The effect of Americanization on the earnings of foreign-born men." Journal of Political Economy 86, 5: 897-921.

CIBULKA, J. (1982) "The new fiscal politics in urban education." California Journal of Teacher Education 9 (Winter).

CLARK, D. L. (1983) "Los Angeles: improbable Los Angeles," in R. M. Bernard and B. R. Rice (eds.) Sunbelt Cities: Politics and Growth Since World War II. Austin: University of Texas Press.

CLARK, T. N. (1975) "The Irish ethic and the spirit of patronage." Ethnicity 2: 305-359.

CLARKE, G. J. (1979) "In defense of deinstitutionalization." Milbank Memorial Fund Quarterly 57, 4: 461-479.

CLOWARD, R. A. and F. F. PIVEN (1974) The Politics of Turmoil. New York: Pantheon.

COBB, C. W. (1972) "Community mental health services and the lower socioeconomic class: a summary of research literature on outpatient treatment (1963-1969)." American Journal of Orthopsychiatry 42: 404-414.

Coca-Cola U.S.A. (1983) Hispanic Business Agenda. Atlanta: Author.

COLBURN, D. R. and G. E. POZZETA (1979) America and the New Ethnicity. Port Washington, NY: Kennikat.

COLE, J. and M. PILISUK (1976) "Differences in the provisions of mental health services by race," American Journal of Orthopsychiatry 46: 510-525.

COLE, N. J., C.H.H. BRANCH, and M. ORLA (1957) "Mental illness," AMA Archives of Neurology and Psychiatry 77: 393-398.

COMAS-DIAZ, L. (1981) "Puerto Rican espiritismo and psychotherapy," American Journal of Orthopsychiatry 51,(4): 636-645.

COMMONS, J. R. (1907) Races and Immigrants in America. New York: Macmillan.

Community Analysis Bureau, Los Angeles (1977) An Ethnic Trend Analysis of Los Angeles County: 1950-1980. Los Angeles: Author.

Community Council of Greater New York (1979) "New immigrants in New York City. Summaries of research in progress and recent reports." (mimeo)

COMSTOCK, G. W. and K. J. HELSING (1976) "Symptoms of depression in two communities," Psychological Medicine 6: 551-563.

CONK, M. A. (1984) "The census, political power and social change." Social Science History 8: 81-106.

CONNABLE, A. and E. SILBERFARB (1967) Tigers of Tammany: Nine Men Who Ran New York City. New York: Holt, Rinehart & Winston.

CONNORS, R. J. (1971) A Cycle of Power: The Career of Jersey City Major Frank Hague. Metuchen, NJ: Scarecrow.

CORCORAN, M. et al. (1980) "Information and Influence Networks in Labor Markets," in G. Duncan et al. (eds.) 5,000 American Families: Patterns of Economic Progress, vol. 8. Ann Arbor, MI: Institute for Social Research.

CORDASCO, F. [ed.] (1968) Jacob Riis Revisited. Garden City, NY: Doubleday.

CORNELIUS, W. A. (1979) "Mexican and Caribbean migration to the United States: the state of current knowledge and recommendations for future research." Working paper 2. Center for U.S.-Mexican Studies, University of California, San Diego.

CORTEZ, A. (1981) "The Texas Assessment of Basic Skills: an analysis of findings." Newsletter (November). San Antonio, TX: Intercultural Development Research Association.

Council of the Great City Schools (1983) Statistical Profile of the Great City Schools: 1970-1982. Washington, DC: Author.

CRAIN, R. L. (1978) The Crawford Remedy. Report to the Superior Court of California for the County of Los Angeles. Case 822854. Los Angeles: Superior Court of California.

CREWDSON, J. (1983) The Tarnished Door: The New Immigrants and the Transformation of America. New York: Time Books.

CRUSE, H. (1967) The Crisis of the Negro Intellectual: From the Origins to the Present. New York: Morrow.

CRUZ, W. (1981) "Board still crowds minority pupils into mobile classrooms." Chicago Reporter (December).

CUFF, D. F. (1984) "A changing economy keeps bankruptcies high." New York Times (March 25).

DAHL, R. A. (1961) Who Governs? New Haven, CT: Yale University Press.

DAVIS, A. F. (1967) Spearheads for Reform: The Social Settlements and the Progressive Movement, 1880-1914. New York: Oxford University Press.

DAVIS, D. B. (1975) The Problem of Slavery in the Age of Revolution, 1770-1823. Ithaca, NY: Cornell University Press.

DE ANDA, D. (1984) "Bicultural socialization: factors affecting the minority experience." Social Work 29(March-April): 101-108.

DE TOCQUEVILLE, A. (1956) Democracy in America. New York: Mentor.

DOHRENWEND, B. P. and B. S. DOHRENWEND (1969) Social Status and Psychological Disorder: A Causal Inquiry. New York: John Wiley.

————M. S. GOULD, B. LILNK, R. NEUGEBAUER, and R. WUNSCH-HITZIG (1980) Mental Illness in the United States. New York: Praeger.

DOHRENWEND, B. P., G. EGRI, and F. S. MENDELSOHN (1971) "Psychiatric disorder in general populations: a study of the problem of clinical judgment." American Journal of Psychiatry 127: 1304-1312.

DORF, I. (1984) "Arthur Imperatore's possible dream." Monthly New Jersey 9, 9: 39-43, 102-103.

DRUCKER, P. (1954) The Practice of Management. New York: Harper & Row.

DUNCAN, O. D. (1969) "Inheritance of poverty or inheritance of race?" in D. P. Moynihan (ed.) Understanding Poverty. New York: Basic Books.

EATON, J. W. and R. S. WEIL (1955) Culture and Mental Disorders. New York: Free Press.

EATON, W. and L. KESSLER (1981) "Rates of symptoms of depression in a national sample." American Journal of Epidemiology 114: 528-538.

EDGERTON, J. W., W. BENTZ, and W. HOLLISTER (1970) "Demographic factors and responses to stress among rural people." American Journal of Public Health 60: 1965-1971.

EDWARDS, P. K. (1981) Strikes in the United States, 1881-1974. New York: St. Martin's.

EGAWA, J. and N. TASHIMA (1982) "Indigenous healers in southeast Asian refugee communities." Pacific Asian Mental Health Research Project, San Francisco. (unpublished)

EITZEN, D. S. and D. A. TIMMER (1984) Criminology. New York: John Wiley.

ERIE, S. P. (1984a) "Building the new machines, 1896-1928: The Irish versus the new immigrants." University of California, San Diego. (unpublished)

————(1984b) "The Depression, New Deal and changing Irish machine fortunes, 1928-1950. "Presented at the annual meeting of the Western Political Science Association.

————(1983) "The organization of Irish-Americans into urban institutions: building the political machine, 1840-1896." Presented at the annual meeting of the American Political Science Association.

————(1980) "Two faces of ethnic power: comparing the Irish and Black experiences." Polity 13: 261-284.

————(1978) "Politics, the public sector and Irish social mobility: San Francisco, 1970-1900." Western Political Quarterly 31: 274-289.

————and M. K. BROWN (1979) "Social policy and the emergence of the Black social welfare economy." Presented at National Conference for Public Administration, Baltimore.

ESPINOSA, R. (1982) Report on the Los Angeles Unified School District: A Comparison of School Finance and Facilities Between Hispanic and Non-Hispanic Schools During Fiscal Year 1980-1981. Los Angeles: Mexican American Legal Defense and Education Fund.

FARLEY, R. (1979) Changes in School Integration and White Enrollment: Fall 1978. Case 822854. Supplementary Report to the Superior Court of California for the County of Los Angeles.

FERNANDEZ, R. R. and J. T. GUSKIN (1981) "Hispanic students and school desegregation," in W. D. Hawley (ed.) Effective School Desegregation: Equity, Quality, and Feasibility. Beverly Hills, CA: Sage.

FINK, L. (1983) Workingmen's Democracy: The Knights of Labor and American Politics. Urbana: University of Illinois Press.

FINK, P. J. and S. P EINSTEIN (1979) "Whatever happened to psychiatry? The deprofessionalization of community mental health centers." American Journal of Psychiatry 136, 4A(April): 406-409.

FLAHERTY, J. A. and R. MEAGHER (1980) "Measuring racial bias in inpatient treatment." American Journal of Psychiatry 137: 679-682.

FOGEL, R. and S. ENGERMANN (1974) Time on the Cross. Boston: Little, Brown.

FONER, E. (1970) Free Soil, Free Labor, Free Men: The Ideology of the Republican Party before the Civil War. New York: Oxford University Press.

FONER, N. (1979) "West Indians in New York City and London: a comparative analysis." International Migration Review 13, 2: 284-297.

Ford Foundation (1984) "Hispanics: challenges and opportunities." Working paper, June.

FREDERICKSON, G. (1971) The Black Image in the White Mind. New York: Harper & Row.

FREEDMAN, M. (1982) "The structure of the labor market and associated training patterns," in Education and Work 1982. Chicago: University of Chicago Press.

———(1976) Labor Markets: Segments and Shelters. Montclair, NJ: Allanhead, Osmun.

———and J. KORAZIM (in press) "Israelis in the New York Area labor force." Contemporary Jewry.

———(in press) "Training for telecommunications occupations." Report prepared for the Center for Advanced Technology in Telecommunications, Polytechnic Institute of New York.

FRERICHS, R. R., C. S. ANESHENSEL, V. A. CLARK (1981) "Prevalence of depression in Los Angeles County." American Journal of Epidemiology 1113: 691-699.

FRIEDLANDER, W. A. (1961) Introduction to Social Welfare. Englewood Cliffs, NJ: Prentice-Hall.

FRIEDMAN, M. (1972) Capitalism and Freedom. Chicago: University of Chicago Press.

FUCHS, L. H. (1983) "Immigration, pluralism, and public policy: the challenge of the pluribus to the unum," in M. M. Kritz (ed.) U.S. Immigration and Refugee Policy: Global and Domestic Issues. Lexington, MA: D. C. Heath.

———(1950) The Political Behavior of American Jews. New York: Free Press.

GALLO, C. (1983) "The construction industry in New York City: immigrants and Black entrepreneurs." New York: Conservation of Human Resources, Columbia University.

GARCIA, F. C. and R. de la GARZA (1982) The Chicano Political Experience: Three Perspectives. North Scituate, MA: Duxbury.

GARCIA, J. R. (1980) Operation Wetback: The Mass Deportation of Mexican Undocumented Workers in 1954. Westport, CT: Greenwood.

GARCIA, P. (1985) Immigration and Language Issues in Urban Ecology: The Case of the Los Angeles Metropolitan Area. Technical Report. Los Alamitos, CA: National Center for Bilingual Research.

GARIS, R. (1927) Immigration Restriction. New York: Macmillan.

GARREAU, J. (1981) The Nine Nations of North America. Boston: Houghton Mifflin.

GARRETT, C. (1961) The La Guardia Years: Machine and Reform Politics in New York City. New Brunswick, NJ: Rutgers University Press.

GARRISON, V. and C. I. WEISS (1979) "Dominican family networks and United States immigration policy: a case study." International Migration Review 13, 2: 264-283.

GAVIT, J. P. (1971) Americans by Choice. Montclair, NJ: Patterson Smith.

GENEN, A. (1971) John Kelly: New York's First Irish Boss. Ph.D. dissertation, New York University.

GIDDENS, A. (1973) The Class Structure of Advanced Societies. New York: Harper & Row.

GILLETTE, W. (1979) Retreat from Reconstruction, 1868-1879. Baton Rouge: Louisiana State University Press.

GILLIN, J. L. (1935) Crime and Penology. New York: Appleton-Century.

GLAAB, C. N. and A. T. BROWN (1967) A History of Urban America. New York: Macmillan.

GLASS, G. V. and M. L. SMITH (1978) Meta-analysis of Research on the Relationship of Class-Size and Achievement. San Francisco: Far West Laboratory.

GLAZER, N. and D. P. MOYNIHAN (1964) Beyond the Melting Pot. Cambridge: MIT Press.

GOFFMAN, E. (1961) Asylums: Essays on the Social Situation of Mental Patients and Other Inmates. Garden City, NY: Doubleday.

GOLD, H. and F. R. SCARPITTI [eds.] (1967) Combatting Social Problems: Techniques of Intervention. New York: Holt, Rinehart & Winston.

GOLAN, N. and R. GRUSCHKA (1971) "Integrating the new immigrant: a model for social work practice in transitional states." Social Work 16, 2: 82-87.

GORDON, D., R. EDWARDS, and M. REICH (1982) Segmented Work, Divided Workers. New York: Cambridge University Press.

GOSNELL, H. F. (1968) Machine Politics: Chicago Model. Chicago: University of Chicago Press.

———(1935) Negro Politicians: The Rise of Negro Politics in Chicago. Chicago: University of Chicago Press.

GOTTFRIED, A. (1962) Boss Cermak of Chicago. Seattle: University of Washington Press.

GRANOVETTER, M. (1974) Getting a Job. Cambridge, MA: Harvard University Press.

GRANTHAM, R. J. (1973) "Effects of counselor, sex, race and language style on black students in initial interviews." Journal of Counseling Psychology 20: 539-559.

GREBLER, L., J. W. MOORE, and R. GUZMAN (1970) The Mexican American People. New York: Free Press.

GREELEY, A. (1976) Ethnicity, Denomination, and Inequality. Beverly Hills, CA: Sage.

———(1972) That Most Distressful Nation: The Taming of the American Irish. Chicago: Quadrangle.

———(1969) Why Can't They be Like Us? Facts and Fallacies About Ethnic Differences and Group Conflicts in America. New York: Institute of Human Relations Press.

GREENSTONE, J. D. and P. E. PETERSON (1973) Race and Authority in Urban Politics: Community Participation and the War on Poverty. New York: Russell Sage Foundation.

GREENWOOD, M. J. (1983) "Regional economic aspects of immigrant location patterns in the United States," in M. Kritz (ed.) U.S. Immigration and Refugee Policy: Global and Domestic Issues. Lexington, MA: D. C. Heath.

GREER, C. (1972) The Great School Legend. New York: Penguin.

GURAK, D. T. and M. M. KRITZ (1983a) "Dominican and Colombian women in New York City: household structure and employment patterns." Migration Today 10, 3/4: 16-21.
———(1983b) "Kinship networks and the settlement process: Dominican and Colombian immigrants in New York City." Presented at "La inmigracion Dominicana en los Estado Unidos," El Museo del Hombre Dominicano/New York Research Program in Inter-American Affairs, Santo Domingo, April.
GURIN, G., J. VEROFF, and S. FELD (1960) Americans View Their Mental Health. New York: Basic Books.
HALPERN, R. (1973) Survival: Black/White. New York: Pergamon.
HAMILTON, C. V. (1976) "Public policy and some political consequences," in M. R. Barnett and J. A. Hefner (eds.) Public Policy for the Black Community. New York: Alfred.
———(1974) "Blacks and the crisis of political participation." Public Interest 34: 188-210.
HANDLIN, O. (1957) Race and Nationality in American Life. Garden City, NY: Doubleday.
HARRING, S. (1976) "The development of the police in the United States." Crime and Social Justice 5 (Spring/Summer).
HARVEY, P. (1970) "Problems in Chinatown." Human Events 30 (May 16): 21.
HENDERSON, T. M. (1976) Tammany Hall and the New Immigrants. New York: Arno.
HERNANDEZ, J. (1983) Puerto Rican Youth Employment. Maplewood, NJ: Waterfront.
HIGGS, R. (1982) "Accumulation of property by southern Blacks before World War I." American Economic Review 72 (September): 725-737.
HIGHAM, J. (1967) Strangers in the Land: Patterns of American Nativism, 1860-1925. New York: Atheneum.
HIRSCH, A. R. (1983) Making the Second Ghetto: Race and Housing in Chicago. 1940-1960. New York: Cambridge University Press.
HIRSCHMAN, C. and M. G. WONG (in press) "Socioeconomic gains of Asian-Americans, Blacks, and Hispanics: 1960-1976." American Journal of Sociology.
———(1983) "Immigration, education and Asian-Americans: a cohort analysis." Presented at the 1983 annual meeting of the American Sociological Association, Detroit.
———(1981) "Trends in socioeconomic achievement among immigrant and native-born Asian-Americans." Sociological Quarterly 22: 495-513.
HOFSTADTER, R. (1944) Social Darwinism in American Thought. Boston: Beacon.
HOOVER, K. (1979) "Organizational networks and ethnic persistence: a case study of Norwegian-American ethnicity in the New York City metropolitan area." Ph.D. dissertation, City University of New York.
HORSMAN, R. (1981) Race and Manifest Destiny. Cambridge, MA: Harvard University Press.
HOWE, I. (1976) World of Our Fathers. New York: Simon & Schuster.
HOWE, M. (1984) "Haitians quietly find better life in the city, despite their fears." New York Times (May 12).
HUNT, J. (1973) "Rapists have big ears: genetic screening in Massachusetts." Real Paper (July 4): 4.
HUTHMACHER, J. J. (1962) "Urban liberalism and the age of reform." Mississippi Valley Historical Review 49: 231-241.

IANNNI, F. J. (1974) Black Mafia: Ethnic Succession in Organized Crime. New York: Simon & Schuster.

ILFELD, R. (1978) "Psychological status of community residents along major demographic dimensions." Archives of General Psychiatry 35: 716-724.

JACKSON, J. (1979) "Illegal aliens: threat to Black workers." Ebony 34: 33-36.

JEFFREY, C. R. (1978) "Criminology as an interdisciplinary science." Criminology 16: 149-170.

JENKINS, S. (1983) "Children who are newcomers: social service needs," in M. Frank (ed.) Newcomers to the United States: Children and Families. New York: Haworth.

JIOBU, R. M. (1976) "Earnings differentials between whites and ethnic minorities: the cases of Asian Americans, Blacks, and Chicanos." Sociology and Social Research 61: 24-38.

JOHNSON, D. R. (1983) "San Antonio: the vicissitudes of boosterism," in R. M. Bernard and B. R. Rice (eds.) Sunbelt Cities: Politics and Growth Since World War Two. Austin: University of Texas Press.

JORDAN, W. (1968) White Over Black. Chapel Hill: University of North Carolina Press.

KAMMERMAN, S. B. and A. J. KAHN (1976) Social Services in the United States: Policies and Programs. Philadelphia: Temple University Press.

KANTROWITZ, N. (1973) Ethnic and Racial Segregation in the New York Metropolis. New York: Praeger.

KARNO, M. and R. B. EDGERTON (1969) "Perception of mental illness in a Mexican-American community." Archives of General Psychiatry 20: 233-238.

KASARDA, J. D. (1984) "Hispanics and city change." American Demographics (November): 24-29.

KATZ, S. N. and J. M. MURRIN (1983) Colonial America. New York: Knopf.

KATZNELSON, I. (1976) "The crisis of the capitalist city: urban politics and social control," in W. D. Hawley and M. Lipsky (eds.) Theoretical Perspectives on Urban Politics. Englewood Cliffs, NJ: Prentice-Hall.

KAUFFMAN, L. (1984) Telephone conversation with attorney for the plaintiff in *Edgewood v. Bynum*, December 11.

KEELEY, C. (1979) U.S. Immigration: A Policy Analysis. New York: Population Council.

——(1975a) "Effects of U.S. Immigration laws on manpower characteristics of immigrants." Demography 12: 179-192.

——(1975b) "Immigration composition and population policy," in P. Reining and I. Tinker (eds.) Population: Dynamics, Ethics, and Policy. Washington, DC: American Association for the Advancement of Science.

——(1974) "The demographic effects of immigration legislation and procedures." Interpreter Releases 51: 89-93.

——(1971) "Effects of the Immigration Act of 1965 on selected population characteristics of immigrants to the U.S." Demography 8: 157-169.

KELLER, M. (1977) Affairs of State. Cambridge, MA: Harvard University Press.

KELLEY, R. (1974) The Cultural Patterns in American Politics. New York: A. A. Knopf.

KERSON, L. S. (1978) "The social work relationship: a form of gift exchange." Social Work 23: 18-27.

KESSNER, T. (1977) The Golden Door: Italian and Jewish Immigrant Mobility in New York City, 1880-1915. New York: Oxford University Press.

KIEFER, M. (1984) "New faces, old dreams." Chicago (March): 127-135.

KILLINGSWORTH, M. R. (1983) "Effects of immigration into the United States on the U.S. labor market: Analytical and policy isues," in M. M. Kritz (ed.) U.S. Immigration and Refugee Policy: Global and Domestic Issues. Lexington, MA: D. C. Heath.

KILSON, M. (1971) "Political change in the Negro ghetto, 1900-1940s," in N. I. Huggins et al. (eds.) Key Issues in the Afro-American Experience. New York: Harcourt Brace Jovanovich.

KIM, I. (1981) New Urban Immigrants: The Korean Community in New York. Princeton, NJ: Princeton University Press.

———(1979) "Immigrants to urban America: the Korean community in the New York metropolitan area." Ph.D. dissertation, City University of New York.

KINCAID, J. (1981) "Political success and policy failure: the persistence of machine politics in Jersey City." Ph.D. dissertation, Temple University.

KITANO, H. (1969) "Japanese-American mental illness," in S. Plog and R. Edgerton (eds.) Changing Perspectives in Mental Illness. New York: Holt, Rinehart & Winston.

———(1967) "Japanese-American crime and delinquency." Journal of Psychology 16: 149-170.

KLEIN, D. P. (1984) "Occupational employment statistics for 1972-1982." Employment and Earnings 31(January): 13-16.

KLEIN, L. Y. (1969) "Some factors in the psychiatric treatment of Spanish-Americans." American Journal of Psychiatry 125: 1674-1681.

KLERMAN, G. L. (1977) "Better but not well: social and ethnical issues in the deinstitutionalization of the mentally ill." Schizophrenia Bulletin 3, 4: 617-631.

KOH, S. D. et al. (1984) "Mental health and relocation stress in Asian elderly immigrants." Pacific/Asian American Mental Health Research Center, Chicago. (unpublished)

KOLM, R. (1971) "Ethnicity and society and community," in O. Feinstein (ed.) Ethnic Groups in the City: Culture, Institutions, and Power. Lexington, MA: D. C. Heath.

KORNBLUM, W. (1974) Blue Collar Community. Chicago: University of Chicago Press.

KRITZ, M. M. [ed.] (1983) U.S. Immigration and Refugee Policy: Global and Domestic Issues. Lexington, MA: D. C. Heath.

———et al. [ed.] (1981) Global Trends in Migration: Theory and Research on International Population Movements. New York: Center for Migration Studies.

KUO, W. (1984) "Prevalence of depression among Asian-Americans." Journal of Nervous and Mental Disease 172, 8: 449-457.

KYLE, C. (1984) Hispanic Drop Out Problem. Co-sponsored by Aspira Inc. of Illinois, National Center for Bilingual Research, Hispanic Policy Development Project.

LAIRD, J. (1984) "Sorcerers, shamans, and social workers: the use of ritual in social work practice." Social Work 29(March-April): 123-129.

LANE, R. (1971) Policing the City: Boston, 1822-1885. New York: Atheneum.

LANGNER, T. S. and S. T. MICHAEL (1963) Life Stress and Mental Health. New York: Free Press.

LARSON, C. J. (1984) Crime-Justice and Society. Bayside, NY: General Hall.

Latino Institute (1984) "A survey of Latino households in Chicago." Lider 3(January): 1-7.

———(1983a) Latinos in Metropolitan Chicago: A Study of Housing and Employment. Monograph Series 6. Chicago: Author.

———(1983b) "Chicago's Latino population grows by 100,000 from 1980 to 1983." Lider 1(July): 1-2.

LEUCHTENBURG, W. (1958) The Perils of Prosperity. Chicago: University of Chicago Press.

LIEBERSON, S. (1980) A Piece of the Pie. Berkeley: University of California Press.
———(1963) Ethnic Patterns in American Cities. New York: Free Press.
LIN, K. M., T. S. INUI, A. M. KLEINMAN, and W. M. WOMACK (1982) "Sociocultural determinants of the help-seeking behavior of patients with mental illness." Journal of Nervous and Mental Disease 170: 78-85.
LIU, W. T. (in press) "Conceptual problems in refugee research," in M. Owan et al. (eds.) Sourcebook of Southeast Asian Refugees. Washington, DC: Government Printing Office.
———(1979) Transition to Nowhere. Nashville: Charter House. (Distributed by the Pacific/Asian American Mental Health Research Center, Chicago.)
LLANES, J. (1982) Cuban Americans: Masters of Survival. Cambridge, MA: Abt Books.
LOGAN, J. R. and L. B. STERNS (1981) "Suburban racial segregation as a nonecological process." Social Forces 60: 61-72.
LOPEZ, A. (1982) "Los Angeles County most ghettoized in Southern California." Los Angeles Times (June 14).
———(1974) "The Puerto Rican diaspora: a survey," in A. Lopez and J. Petras, Puerto Rico and Puerto Ricans: Studies in History and Society. New York: John Wiley.
LORION, R. P. (1978) "Research on psychotherapy and behavior change with the disadvantaged: past, present and future directions," in S. L. Garfield and A. E. Bergin (eds.) Handbook on Psychotherapy and Behavior Change: An Empirical Analysis. New York: John Wiley.
———(1974) "Patient and therapist variables in the treatment of low-income patients." Psychological Bulletin 81: 344-354.
———(1973) "Socioeconomic status and traditional treatment approaches reconsidered." Psychological Bulletin 81: 344-354.
LOWI, T. J. (1964) At the Pleasure of the Major. New York: Free Press.
LUBELL, S. (1965) The Future of American Politics. New York: Harper & Row.
LUBORSKY, L., M. CHANDLER, A. H. AVERBACH, J. COHEN, and H. M. BACHRACH (1971) "Factors influencing the outcome of psychotherapy: a review of quantitative research." Psychological Bulletin 75: 145-185.
LUCAS, I. (1971) Puerto Rican Dropouts in Chicago: Numbers and Motivation. Final report to the Office of Education (Project O-E-108).
MAGILL, R. (1984) Social Policy in American Society. New York: Human Sciences Press.
———(1979) Community Decision Making for Social Welfare: Federalism, City Government and the Poor. New York: Human Sciences Press.
MALDONADO-DENIS, M. (1980) The Emigration Dialectic: Puerto Ricans and the USA. New York: International Publishers.
MANIS, J. G., M. J. BRAWER, C. L HUNT, and L. C. KERCHER (1964) "Estimating the prevalence of mental illness." American Sociological Review 29: 84-89.
MANN, A. (1965) La Guardia Comes to Power, 1933. New York: J. B. Lippincott.
MANN, E. S. and J. J. SALVO (1984) "Characteristics of new Hispanics." Presented at the annual meeting of the Population Association of America, Minneapolis.
MARABLE, M. (1983) "How Washington won: the political economy of race in Chicago." Intergroup Relations 11: 56-81.
MARCOS, L. R. (1979) "Effects of interpreters on the evaluation of psychopathology in non-English-speaking patients." American Journal of Psychiatry 136: 171-174.
MARMOR, T. (1973) The Politics of Medicare. Chicago: Aldine.

MARSH, R. E. (1980) "Socioeconomic status of Indo-Chinese refugees in the United States: Progress and problems." Social Security Bulletin 43, 10: 11-20.

MASSEY, D. S. (1983) "A research note on residential succession: the Hispanic case." Social Forces 61: 825-833.

———(1981a) "Dimensions of the new immigration to the United States and the prospects for assimilation." Annual Review of Sociology 7: 57-85.

———(1981b) "Hispanic residential segregation: a comparison of Mexicans, Cubans, and Puerto Ricans." Sociology and Social Research 65: 311-322.

———(1979a) "Residential segregation of Spanish Americans in the U.S. urban areas." Demography 8: 553-564.

———(1979b) "Effects of socioeconomic factors on the residential segregation of Black and Spanish Americans in U.S. urbanized areas." American Sociological Review 44: 1015-1022.

———(1978) "On the measurement of segregation as a random variable." American Sociological Review 43: 587-590.

MATTHEWS, F. H. (1977) Quest for an American Sociology: Robert E. Park and the Chicago School. Montreal and London: McGill Queens University Press.

MAYO, J. A. (1974) "Utilization of a community mental health center by blacks: Admission to inpatient status." Journal of Nervous and Mental Disease 158: 202-7.

MAYO-SMITH, R. (1890) Emigration and Immigration. New York: Charles Schribners.

McCARTHY, J. D. and W. L. YANCEY (1971) "Uncle Tom and Mr. Charlie: metaphysical pathos in the study of racism and personal disorganization." American Journal of Sociology 76: 648-72.

McDONALD, R. J. (1984) "The 'underground economy' and BLS statistical data." Monthly Labor Review 107 (January): 4-18.

McKENZIE, R. D. (1967) "The ecological approach to the study of the human community," in R. E. Park and E. W. Burgess (eds.) The City. Chicago: Aldine.

McMILAN, A. W. and A. K. BIXBY (1980) "Social welfare expenditures, fiscal year, 1978." Social Security Bulletin 43 (May): 3-17.

McPHERSON, J. (1982) Ordeal by Fire: The Civil War and Reconstruction. New York: A. A. Knopf.

MEDNICK, S. and V. VOLAVAKA (1980) "Biology and crime," in N. Morris and M. Tonry (eds.) Biosocial Bases of Criminal Behavior. Chicago: University of Chicago Press.

MEIER, A. and E. RUDWICK (1970) From Plantation to Ghetto. New York: Hill & Wang.

MEILE, R. L. and P. N. HAESE (1969) "Social status, status incongruence and symptoms of stress." Journal of Health and Social Behavior 10: 237-244.

MEKETON, M. J. (1983) "Indian mental health: an orientation." American Journal of Orthopsychiatry 53, 1: 110-115.

MERTON, R. K. (1968) Social Theory and Social Structure. New York: Free Press.

MEYER, C. H. (1984) "Working with new immigrants." Social Work 29 (March/April): 99.

MEYER, S. (1981) The Five Dollar Day. Albany: SUNY Press.

MILLER, H. P. (1966) Income Distribution in the United States. Washington, DC: Government Printing Office.

MILLER, P. (1956) Errand into the Wilderness. Cambridge, MA: Harvard University Press.

MILLER, W. B. (1958) "Lower class culture as a generating milieu of gang delinquency." Journal of Social Issues 14: 5-19.

MILLS, C. W. (1959) The Sociological Imagination. New York: Oxford University Press.

MIROWSKY, J. and C. ROSS (1980) "Minority status, ethnic culture, and distress: a comparison of blacks, whites, Mexicans, and Mexican Americans." American Journal of Sociology 86: 479-495.

MIRRINGOFF, M. L. (1980) "The impact of public policy upon social welfare." Social Service Review 54 (September): 301-316.

MOHL, R. A. (1983) "Miami: the ethnic cauldron," in R. M. Bernard and B. R. Rice (eds.) Sunbelt Cities: Politics and Growth Since World War Two. Austin: University of Texas Press.

MOLLENKOPF, J. H. (1983) The Contested City. Princeton, NJ: Princeton University Press.

MOLLICA, R. F. (1983) "From asylum to community: the threatened disintegration of public psychiatry." New England Journal of Medicine 308, 7: 367-383.

———(1980a) "Community mental health centres: an American response to Kathleen Jones." Journal of Research in Social Medicine 73: 863-870.

———(1980b) "Equity and changing patient characteristics—1950-1975." Archives of General Psychiatry 37: 1257-1263.

———J. D. BLUM, and F. C. REDLICH (1980) "Equity and the psychiatric care of the Black patient: 1950-1975." Journal of Nervous and Mental Disease 168: 279-286.

MONTERO, D. (1979a) Vietnamese Americans: Patterns of Resettlement and Socioeconomic Adaptation in the United States. Boulder, CO: Westview.

———(1979b) "Vietnamese refugees in America: toward a theory of spontaneous international migration." International Migration Review 13, 4: 624-648.

MONTGOMERY, D. (1979) Workers' Control in America. New York: Cambridge University Press.

———(1967) Beyond Equality. New York: Knopf.

MOORE, J. (1981) "Minorities in the American class system." Daedalus 110 (Spring): 275-299.

———(1979) "Mexican Americans," in D. R. Colburn and G. E. Possetta (eds.) America and the New Ethnicity. Port Washington, NY: Kennikat.

———(1976) "Colonialism: the case of the Mexican Americans," in C. E. Cortes et al. (eds.) Three Perspectives on Ethnicity: Blacks, Chicanos, and Native Americans. New York: G. P. Putnam.

———and H. PACHON (1985) Hispanics in the United States. Englewood Cliffs, NJ: Prentice-Hall.

MORAN, B. (1978) "Biomedical research and the politics of crime control: a historical perspective." Contemporary Crises 2.

MORGAN, E. (1975) American Slavery, American Freedom: The Ordeal of Colonial Virginia. New York: Norton.

MORRIS, C. R. (1981) The Cost of Good Intentions: New York City and the Liberal Experiment, 1960-1975. New York: McGraw-Hill.

MOSCOW, W. (1971) The Last of the Big-Time Bosses: The Life and Times of Carmine De Sapio and the Rise and Fall of Tammany Hall. New York: Stein & Day.

MULLER, T. (1984) The Fourth Wave. Washington, DC: Urban Institute Press.

MURRAY, R. K. (1955) Red Scare. Minneapolis: University of Minnesota Press.

MYERS, B. (1971) History of Tammany Hall. New York: Dover.

MYERS, J. K. and M. N. WEISSMAN (1980) "Screening for depression in a community sample: the use of a self-report symptom scale to detect the depressive syndrome." American Journal of Psychiatry 137: 1081-1084.

MYERS, J. K. et al. (1984) "Six-month prevalence of psychiatric disorders in three communities." Archives of General Psychiatry 41, 10: 959-970.

National Assessment of Educational Progress (1982a) Performance of Hispanic Students in Two National Assessments of Reading. Denver: Education Commission of the States.

———(1982b) Students from Hmong in Which English Is Not the Dominant Language: Who Are They and How Well Do They Read? Denver: Education Commission of the States.

National Center for Educational Statistics (1981) Contractor Report, Achievement of Hispanic Students in American High Schools: Background Characteristics and Achievement. Washington, DC: Government Printing Office.

———(1980) High School and Beyond. Washington, DC: Government Printing Office.

National Commission on Excellence in Education (1983) A Nation at Risk. Washington, DC: U.S. Department of Education.

National Institute of Mental Health (1978) "Mental disorders," in Health United States, 1978. DHEW Publication (PHS) 78-1232. Washington, DC: Department of Health, Education and Welfare.

———(1976) Provisional Data on Federally Funded Community Mental Health Centers, 1974-1975. Rockville, MD: NIMH, Division of Biometry and Epidemiology, Survey and Reports Branch.

———(1971) Community Mental Health Center Program Operating Handbook, vol. I. Policy and Standards Manual. Washington, DC: Department of Health, Education and Welfare.

NEFF, J. A. and B. A. HUSAINI (1980) "Race, socioeconomic status, and psychiatric impairment: a research note." Journal of Community Psychology 8: 16-19.

NELSON, D. (1980) Frederick W. Taylor and the Rise of Scientific Management. Madison: University of Wisconsin Press.

———(1975) Managers and Workers. Madison: University of Wisconsin Press.

NORTH, D. (1983) "Impact of legal, illegal, and refugee migration on U.S. social service programs," in M. M Kritz (ed.) U.S. Immigration and Refugee Policy: Global and Domestic Issues. Lexington: MA: D. C. Heath.

———(1974) Immigrants and the American Labor Market. Manpower Research Monograph 31. Washington, DC: U.S. Department of Labor.

NUGENT, W. (1981) Structures of American Social History. Bloomington: Indiana University Press.

O'CONNOR, L. (1977) Requiem: The Decline and Demise of Mayor Daley and His Era. Chicago: Contemporary Books.

———(1975) Clout: Mayor Daley and His City. Chicago: Henry Regnery.

OLEJARAZYK, K. J. (1971) "The cultural impoverishment of immigrants," in O. Feinstein (ed.) Ethnic Groups in the City: Culture, Institutions and Power. Lexington, MA: Heath.

OLIVAS, M. A. and N. ALCINBA (1979) The Dilemma of Access. Washington, DC: Howard University Press.

OLIVER, M. L. and J. H. JOHNSON (in press) "Interethnic conflict in an urban ghetto: the case of Blacks and Latinos," in R. L. Radcliffe (ed.) Research, Social Movements, Conflict and Change. New York: JAI.

OLSON, J. S. (1979) The Ethnic Dimension in American History (vol. 2). New York: St. Martin's.

ORFIELD, G. (1982) Desegregation of Black and Hispanic Students from 1968 to 1980. A report to the subcommittee on Civil and Constitutional Rights of the committee on the Judiciary of the United States House of Representatives. Washington, DC: Joint Center for Political Studies.
————et al. (1984) The Chicago Study of Access and Choice in Higher Education. A report to the Illinois Senate Committee on Higher Education. Chicago: Committee on Public Policy Studies, University of Chicago.
————(1983) School Desegregation Patterns in the States, Large Cities and Metropolitan Areas, 1968-1980. Washington, DC: Joint Center for Political Studies.
Organization of Chinese Americans (1984) Where We Stand in America. Washington, DC: Author.
ORUM, L. S. (1984a) Hispanic Dropouts: Community Responses. Washington, DC: National Council of La Raza.
————(1984b) Selected Statistics on the Education of Hispanics. Washington, DC: National Council of La Raza.
PADILLA, A., R. A. RUIZ, and R. ALVAREZ (1975) "Community mental health services for the Spanish-speaking/surnamed population." American Psychologist 30, 9: 892-905.
PADILLA, F. M. (1985) Latino Consciousness: The Case of Mexican Americans and Puerto Ricans in Chicago. South Bend: Indiana University of Notre Dame Press.
PAO-MIN, C. (1981) "Health and crime among Chinese-Americans: recent trends." Phylon 42 (December): 356-368.
PARK, R. E. (1925) The City. Chicago: University of Chicago Press.
————and H. A. MILLER (1969) Old World Traits Transplanted. New York: Arno. (Originally published 1921)
PASAMANICK, B. (1962) "A survey of mental disease in an urban population: an approach to total prevalence by age." Mental Hygiene 46: 567-725.
————D. W. ROBERTS, D. W. LEMKAU, and D. B. KRUEGER (1959) "A survey of mental disease in an urban population: prevalence by race and income," in B. Pasamanick (ed.) Epidemiology of Mental Disorder. Washington, DC: American Association for the Advancement of Science.
PASSELL, J. S. and K. A. WOODROW (1984) "Geographic distribution of undocumented immigrants: estimates of undocumented aliens counted in the 1980 census by state." Presented at the annual meeting of the Population Association of America, Minneapolis.
PIERCE, N. R. and J. HAGSTROM (1983) The Book of America: Inside 50 States Today. New York: W. W. Norton.
PETERSEN, J. (1972) "Thunder Out of Chinatown." National Observer 11 (March 8): 1, 18.
PETERSEN, W. (1980) "Concepts of ethnicity," in S. Thernstrom (ed.) Harvard Encyclopedia of American Ethnic Groups. Cambridge, MA: Harvard University Press.
PHILLIPS, D. C. (1966) "The 'true prevalence' of mental illness in a New England state." Community Mental Health Journal 2 (Spring): 35-40.
PIORE, M. (1979) Birds of Passage. New York: Cambridge University Press.
PIVEN, F. F. and R. A. CLOWARD (1982) The New Class War. New York: Pantheon.
————(1981) "Keeping labor lean and hungry." Nation, November 6: 466-467.
————(1971) Regulating the Poor: The Functions of Public Welfare. New York: Vintage.
POLENBERG, R. (1980) One Nation Divisible. New York: Penguin.

PORTES, A. and R. L. BACH (1980) "Immigrant earnings: Cuban and Mexican immigrants in the United States." International Migration Review 14: 315-341.

POSTON, D. and D. ALVIREZ (1973) "On the cost of being a Mexican American." Social Science Quarterly 53: 695-709.

———and M. TIENDA (1976) "Earnings differences between Anglo and Mexican American male workers in 1960 and 1970: changes in the 'cost' of being Mexican American." Social Science Quarterly 57: 618-631.

POTTER, D. (1976) The Impending Crisis of the South. New York: Harper & Row.

President's Commission on Mental Health (1978a) Report of the Special Population Subpanel on Mental Health of Hispanic Americans, Task Panel Reports, Volume III. Washington, DC: Government Printing Office.

———(1978b) Report of the Special Populations Subpanel on Mental Health of American Indians and Alaska Natives. Washington, DC: Government Printing Office.

———(1978c) Report of the Special Population Subpanel on Mental Health of Asians/Pacific Americans. Washington, DC: Government Printing Office.

PRESTON, M. B. (1982) "Black politics and public policy in Chicago: self interest versus constituent representation," in M. B. Preston et al. (eds.) The New Black Politics: The Search for Political Power. New York: Longman.

PROTESS, D. L. (1974) "Banfield's Chicago revisited: the conditions for and social policy implications of the transformation of a political machine." Social Science Review 48: 184-202.

QUERALT, M. (1984) "Understanding Cuban immigrants: a cultural perspective," in Social Work 29(March-April): 115-121.

QUESADA, C. M. (1976) "Language and communication barriers for health delivery to a minority group." Social Science and Medicine 10: 323-327.

———W. SPEARS, and P. RAMOS (1978) "Interacial depressive symptomatology in the Southwest." Journal of Health and Social Behavior 19: 77-85.

QUIMBY, E. (1976) "Black political development in Bedford-Stuyvesant as reflected in the origin and role of the Bedford-Stuyvesant restoration corporation." Ph.D. dissertation, City University of New York.

RAAB, S. (1984) "Asia crime groups spreading in U.S., Smith tells panel." New York Times (October 24): 1, 16.

RABINOVITZ, F. F. (1978) Trends in the Racial and Ethnic Composition of Los Angeles Unified School District During 1966-1977. Report to the Superior Court of California for the County of Los Angeles. Case 822854.

———(1975) Minorities in Suburbs: The Los Angeles Experience. Cambridge, MA: Joint Center for Urban Studies of MIT and Harvard.

RADLOFF, L. S. (1977) "The CES-D scale: a self-report depression scale for research in the general population." Applied Psychological Measurement 1: 385-401.

RAKOVE, M. L. (1975) Don't Make No Waves, Don't Back No Losers: An Insider's Analysis of the Daley Machine. Bloomington: Indiana University Press.

RAPER, A. F. (1974) Preface to Peasantry. New York: Atheneum.

REDICK, R. W. (1976) Addition Rates to Federally Funded Community Mental Health Centers, United States, 1973. Statistical Note 126, DHEW Publication ADM 76-158. Rockville, MD: National Institute of Mental Health.

REDLICH, R. and S. R. KELLERT (1978) "Trends in American mental health." American Journal of Psychiatry 135: 22-28.

REGIER, D. A. et al. (1984) "The NIMH epidemiologic catchment area program." Archives of General Psychiatry 41, 10: 934-941.

REGIER, D. A., I. D. GOLDBERG, and C. A TAUBE (1978) "The de facto U.S. mental health services system." Archives of General Psychiatry 35 (June): 685-693.

REIMERS, C. W. (1980) Sources of the Wage Gap Between Hispanic and other White Americans. Working Paper 139. Princeton, NJ: Princeton University, Industrial Relations Sections.

RICHARDSON, J. (1970) The New York Police: Colonial Times to 1901. New York: Oxford University Press.

RIMLINGER, G. (1971) Welfare Policy and Industrialization in Europe, America and Russia. New York: John Wiley.

RIVERA, G., Jr., J. J. WANDERER, K. PENIRIAN, and L. VIGIL (1984) "Curanderismo and urban Hispanic women: a study of beliefs on natural, emotional, and supernatural illnesses afflicting children." Presented at the annual meeting of the Midwest Sociological Society, Chicago.

ROBERTS, R. E. (1981) "Prevalence of depressive symptoms among Mexican-Americans." Journal of Nervous and Mental Disease 169: 213-219.

———(1980) "Prevalence of psychological distress among Mexican-Americans." Journal of Health and Social Behavior 21: 134-145.

———J. M. STEVENSON, and L. BRESLOW (1981) "Symptoms of depression among blacks and whites in an urban community." Journal of Nervous and Mental Disease 169, 12: 774-779.

ROBINS, L. N., J. E. HELZER, J. CROUGHAN, and K. S. RATLIFF (1982) "The NIMH diagnostic interview schedule: its history, characteristics, and validity," in J. K. Wing et al. (eds.) The Concept of a Case: Theory and Method in Community Psychiatric Surveys. London: Great McIntyre.

ROBINS, L. N. et al. (1984) "Lifetime prevalence of specific psychiatric disorders in three sites." Archives of General Psychiatry 41, 10: 949-958.

ROBINSON, C. (1980) Special Report: Physical and Emotional Health Care Needs of Indochinese Refugees. Washington, DC: Action Center.

ROMO, R. (1982) East Los Angeles: History of an Urban Barrio. Austin: University of Texas Press.

ROSE, S. J. (1983) Social Stratification in the U.S.. Baltimore, MD: Social Graphics.

ROSEN, H. and F. OLIVAS (1980) The Condition of Education for Hispanic Americans. Washington, DC: National Center for Educational Statistics.

ROSENFELD, A. H. (1976) Psychiatric Education. Prologue to the 1980s: Report of the Conference on Education of Psychiatrists. Washington, DC: American Psychiatric Association.

ROSENFELD, C. (1975) "Job seeking methods of American workers." Monthly Labor Review 98 (August): 39-42.

ROSENTHAL, E. (1975) "The equivalence of United States census data for persons of Russian stock or descent with American Jews: an evaluation." Demography 12: 275-290.

ROSS, A. (1958) "Do we have a new industrial feudalism?" American Economic Review 48 (December): 903-920.

ROSS, E. A. (1922) The Social Trend. New York: Century Company.

ROYKO, M. (1971) Boss: Richard J. Daley of Chicago. New York: Signet.

RYAN, D. P. (1983) Beyond the Ballot Box: A Social History of the Boston Irish, 1847-1917. London: Associated University Presses.

RYAN, W. (1971) Blaming the Victim. New York: Pantheon.

SABIN, J. E. (1975) "Translating despair." American Journal of Psychiatry 132: 197-199.

SALPUKAS, A. (1983) "Trucking's great shakeout." New York Times (December 13).

SAMUELSON, R. J. (1983) "Middle-class media myth." National Journal 15 (December 31): 1673-1678.

SANTIAGO, I. S. (1984) "Dropouts or 'pushouts'? Puerto Ricans and school policies." Presented for the roundtable on Puerto Ricans in the Continental United States, sponsored by the Johnson Foundation, Wingspread, Racine, WI, March 22-24.

SANUA, V. D. (1966) "Sociocultural aspects of psychotherapy and treatment: a view of the literature," in L. E. Abt and B. F. Riess (eds.) Progress in Clinical Psychology, vol. 7. New York: Grune & Stratton.

SASSEN-KOOB, S. (1979) "Formal and informal associations: Dominicans and Columbians in New York." International Migration Review 13, 2: 314-332.

SAWYER, K. (1984) "Organized labor lost big in 1983, and New Year promises little more." Washington Post (January 8).

SAYRE, W. S. and H. KAUFMAN (1960) Governing New York City. New York: Russell Sage.

SCHEPER-HUGHES, N. (1981) "Dilemmas in deinstitutionalization: a view from inner city Boston." Journal of Operational Psychiatry 12: 90-99.

SCHICK, A. (1969) "Systems, politics and systems budgeting." Public Administration Review 28: 546-558.

SCHMIDT, W. E. (1983) "Nigerians suspected of link to wide pattern of white-collar fraud in U.S.." New York Times (April 3): 1.

SCHUBERT, D.S.P. and S. I. MILLER (1980) "Differences between the lower social classes: some new trends." American Journal of Orthopsychiatry 50: 712-717.

SCHWAB, J. J., N. H. McGINNIS, and G. J. WARHEIT (1973) "Social psychiatric impairment: racial comparisons." American Journal of Psychiatry 130: 183-187.

SCHWAB, J. J. and G. J. WARHEIT (1972) "Evaluating southern mental health needs and services." Journal of Florida Medical Association: 17-20.

SCHWARTZ, D. (1984) "Update: on minimum competency testing." Newsnotes, Center for Law and Education, Cambridge, MA (October).

SCOTT, J. C. (1969) "Corruption, machine politics, and political change." American Political Science Review 63: 1142-1158.

SHAPIRO, S. et al. (1984) "Utilization of health and mental health services." Archives of General Psychiatry 41, 10: 971-982.

SHEFTER, M. (1978) "The electoral foundations of the political machine: New York City, 1884-1897," in J. S. Silbey et al. (eds.) The History of American Electoral Behavior. Princeton, NJ: Princeton University Press.

SHERGOLD, P. (1982) Working-Class Life: The American Standard in Comparative Perspective, 1899-1913. Pittsburgh: University of Pittsburgh Press.

SHORE, J. H., J. D. KINZIE, J. L. HAMPSON, and E. M. PATTISON (1973) "Psychiatric epidemiology of an Indian village." Psychiatry 36: 41-57.

SIEGEL, P. M. (1965) "On the cost of being a Negro." Sociological Inquiry 35: 41-57.

SIEGEL, P. (1983) "Competency testing and the national origin minority student." Presented at the National Working Session on New Directions in Bilingual/ Bicultural Education Advocacy, San Antonio, TX.

SILBERMAN, C. (1978) Criminal Violence, Criminal Justice. New York: Random House.

SMITH, M. R. (1977) Harpers Ferry Armory and the New Technology. Ithaca, NY: Cornell University Press.

SOLLENBERGER, R. T. (1968) "Chinese-American child rearing practices and juvenile delinquency." Journal of Social Psychology 74: 13-23.

SOLOMON, B. M. (1956) Ancestors and Immigrants. Cambridge, MA: Harvard University Press.

SPITZER, S. (1981) "The political economy of policing," in D. F. Greenberg (ed.) Crime and Capitalism. Palo Atlo, CA: Mayfield.

———and A. T. SCHULL (1977) "Social control in historical perspective: from private to public responses to crime," in D. F. Greenberg (ed.) Corrections and Punishment. Beverly Hills, CA: Sage.

SROLE, L. et al. (1962) Mental Health in the Metropolis. New York: McGraw-Hill.

STACK, C. B. (1974) All Our Kin. New York: Harper & Row.

STANBACK, T. et al. (1981) Services: The New Economy. Totowa, NJ: Allanhead, Osmun.

STANTON, A. H. and M. S. SCHWARTZ (1954) The Mental Hospital. New York: Basic Books.

State of Hawaii (1970) Statistical Report of the Department of Health. Honolulu: Department of Health.

STEINBERG, L. et al. (1984) "Dropping out among language minority youth." Review of Educational Research (Spring): 113-132.

STEINBERG, S. (1981) The Ethnic Myth. New York: Atheneum.

STERN, M. S. (1977) "Social class and psychiatric treatment of adults in the mental health center." Journal of Health and Social Behavior 18: 317-325.

STOLZENBERG, R. M. (1975) "Education, occupation and wage differences between white and black men." American Journal of Sociology 81: 299-323.

STONE, C. (1968) Black Political Power in America. Indianapolis: Bobbs-Merrill.

SUE, S. (1977) "Community mental health services to minority groups. Some optimism, some pessimism." American Journal of Psychologist 32: 616-624.

———and H. McKINNEY (1975) "Asian-Americans in the community mental health care system." American Journal of Orthopsychiatry 45: 111-118.

SUE, S. and J. K. MORISHIMA (1982) The Mental Health of Asian Americans. San Francisco: Jossey-Bass.

SUE, S. and D. W. SUE (1974) "MMPI comparisons between Asian American and non-Asian students utilizing a student health psychiatric clinic." Journal of Consulting Psychology 21: 423-427.

SULLIVAN, T. A. (1978) "Racial-ethnic differences in labor force participation: an ethnic stratification perspective," in F. D. Bean, and W. P. Frisbie (eds.) The Demography of Racial and Ethnic Groups. New York: Academic Press.

SUTTLES, G. (1972) The Social Construction of Communities. Chicago: University of Chicago Press.

TAEUBER, K. F. and A. F. TAEUBER (1965) Negroes in Cities: Residential Segregation and Neighborhood Change. Chicago: Aldine.

THEODORSON, G. A. (1961) Studies in Human Ecology. Evanston, IL: Row & Peterson.

THOMAS, B. (1961) International Migration and Economic Development. New York: UNESCO.

THOMAS, W. I. and F. ZNANIECKI (1958) The Polish Peasant in Europe and America. Chicago: University of Chicago Press.

THOMPSON, E. III (1984) "Race and the Chicago election." Journal of Ethnic Studies 11: 1-10.

THRASHER, F. M. (1963) The Gang. Chicago: University of Chicago Press.

THUROW, L. C. (1980) The Zero Sum Society: Distribution and the Possibilities for Economic Change. New York: Basic Books.

TIENDA, M. (1983) "Socioeconomic and labor force characteristics of U.S. immigration: Issues and approaches," in M. M. Kritz (ed.) U.S. Immigration and Refugee Policy: Global and Domestic Issues. Lexington, MA: D. C. Heath.

———(1981) "Nationality and income attainment among native and immigrant Hispanics in the U.S.", in M. Tienda (ed.) Hispanic Origin Workers in the U.S. Labor Market: Comparative Analyses of Employment Outcomes. Washington, DC: U.S. Department of Labor, Employment and Training Administration.

———and V. ORTIZ (1984) "Hispanicity and the 1980 census." Presented at the annual meeting of the American Sociological Association, San Antonio, TX, August.

TIMBERLAKE, E. M. and K. O. COOK (1984) "Social work and the Vietnamese refugee." Social Work, 29 (March-April): 108-114.

TISCHLER, G. L., J. E. HENISZ, J. K. MYERS, and P. C. BOSWELL (1975a) "Utilization of mental health services. I. Patienthood and the prevalence of symptomatology in the community." Archives of General Psychiatry 32: 411-415.

———(1975b) "Utilization of mental health services. II. Mediators of service allocations." Archives of General Psychiatry 32: 416-418.

TISCHLER, G. L., J. HENISZ, J. K. MYERS, and V. GARRISON (1972) "Catchmenting and the use of mental health services." American Journal of Public Health 46: 173-186.

TITMUSS, R. (1965) "The rule of redistribution in social policy." Social Security Bulletin (June).

TRACHTENBERG, A. (1982) The Incorporation of America: Culture and Society in the Gilded Age. New York: Hill & Wang.

TRUSSEL, R. E., J. ELINSON, and M. LEVIN (1956) "Comparisons of various methods of estimating the prevalence of chronic disease in a community—the Hunterdon County Study," American Journal of Public Health 46: 173-186.

TSCHETTER, J. and J. LUKASIEWICZ (1983) "Employment changes in construction: secular, cyclical, seasonal," Monthly Labor Review 106 (March): 11-17.

TSENG, W. S., J. F. McDERMOTT, Jr., and T. W. MARETZKI (1974) People and Cultures in Hawaii: An Introduction for Mental Health Workers. Honolulu: Department of Psychiatry, University of Hawaii School of Medicine.

UGALDE, A., F. D. BEAN, and G. CARDENAS (1979) "International migration from the Dominican Republic: findings from a national survey," International Migration Review 13, (2): 235-254.

UMBENBAUER, S. I. and L. L. DeWITTE (1978) "Patient race and social class: attitudes and decisions among three groups of mental health professionals," Comprehensive Psychiatry 19: 509-515.

U.S. Bureau of the Census (1984) Current Population Reports, Series P-25, 945, February. Washington, DC: Government Printing Office.

———(1983a) "America's ancestry—'melting pot,' 'salad bowl,' or 'tapestry'?" U.S. Department of Commerce News (June 1).

———(1983b) Census of Population and Housing: 1980. Public Use Microdata Samples. Technical Documentation. Washington, DC: Government Printing Office.

————(1983c) 1980 Census of Population, Volume 1, Characteristics of the Population. Chapter B, General Population Characteristics, Part 1, U.S. Summary, PC80-1-B1. Washington, DC: Government Printing Office.

————(1983c) 1980 Census of Population, Volume 1, Characteristics of the Population. Chapter B, General Population Characteristics, Part 1, U.S. Summary, PC80-1-B1. Washington, DC: Government Printing Office.

————(1983d) 1980 Census of Population, Volume 1, Characteristics of the Population. Chapter C, Detailed Population Characteristics, Section A: United States. Washington, DC: Government Printing Office.

————(1983e) 1980 Census of Population, Volume 1, Characteristics of the Population. Chapter D, Detailed Population Characteristics. Part 1: U.S. Summary. PC801-1-D1. Washington, DC: Government Printing Office.

————(1983f) 1980 Census of Population, General Social and Economic Characteristics: California. PC80-1-C6.

————(1983g) 1980 Census of Population, General Social and Economic Characteristics: New York. PC80-1-C34. Washington, DC: Government Printing Office.

————(1983h) 1980 Census of Population and Housing, Census Tracts: Los Angeles-Long Beach Standard Metropolitan Statistical Area. PHC80-2-226. Washington, DC: Government Printing Office.

————(1982a) 1980 Census of Population and Housing, Provisional Estimate of Social, Economic, and Housing Characteristics. PHC80-s1-1. Washington, DC: Government Printing Office.

————(1982b) "Coverage of the national population in the 1980 census, by age, sex, and race: preliminary estimates by demographic analysis." Current Population Reports, Special Studies P-23: 115. Washington, DC: Government Printing Office.

————(1978) "Microdata from the survey of income and education." Data Access Descriptions 42 (January). Washington, DC: Government Printing Office.

————(1973) "Characteristics of the population by ethnic origin: March 1972 and 1971." Current Population Reports, Series P-20. Washington, DC: Government Printing Office.

U.S. Congress (1966) "Comprehensive health planning and public health services amendments of 1966." Public Law 89-749. Eighty-ninth Congress, second session, October 18.

U.S. Department of Education (1984a) The Condition of Bilingual Education in the Nation. Report from the Secretary of Education to the President and Congress. Washington, DC: Government Printing Office.

————(1984b) Public Response to a Nation at Risk. Washington, DC: Government Printing Office.

U.S. Department of Justice (1981) Annual Report: Immigration and Naturalization Service. Washington, DC: Government Printing Office.

————(1980) American Prisons and Jails, vol. 7, "Population trends and projections." Washington, DC: National Institute of Justice.

————(1975) Annual Report: Immigration and Naturalization Service. Washington, DC: Government Printing Office.

————(1974) Annual Report: Immigration and Naturalization Service. Washington, DC: Government Printing Office.

————(1973) Annual Report: Immigration and Naturalization Service. Washington, DC: Government Printing Office.

————(1972) Annual Report: Immigration and Naturalization Service. Washington, DC: Government Printing Office.

————(1970) Annual Report: Immigration and Naturalization Service. Washington, DC: Government Printing Office.

U.S. Department of Labor (1982) Employment and Training Report of the President. Table A-23. Washington, DC: Government Printing Office.

U.S. General Accounting Office (1983) Report on Mental Health Manpower. Washington, DC: Government Printing Office.

U.S. Immigration Commission (1911) Immigration and Crime, vol. 36. Washington, DC: Government Printing Office.

U.S. Library of Congress (1980) Selected Readings on U.S. Immigration Policy and Law. Washington, DC: Government Printing Office.

U.S. National Commission on Law Enforcement (1931) Report on Crime and Foreign Born. Washington, DC: Government Printing Office.

U.S. Select Commission on Immigration and Refugee Policy (1981) U.S. Immigration Policy and the National Interest. Staff Report Supplement to the Final Report and Recommendations. Washington, DC: Government Printing Office.

USDAN, M. D. (1984) "New trends in urban demography." Education and Urban Society 16, 4: 399-414.

VALENTINE, B. (1978) Hustling and Other Hard Work. New York: Free Press.

VALLE, J. and A. R. RIESTER (1976) "Utilization pattern of mental health services by the Spanish-speaking/Spanish surnamed population and services rendered by the Community Mental Health Center of South Florida." Presented at the First Annual Southeast Hispanic Conference on Human Services, Miami, February 6.

VAN ARSDOL, M. D. and L. S. SCHUERMAN (1971) "Redistribution and assimilation of ethnic populations." Demography 8: 459-480.

VANDENBOS, G. R. and J. STAPP (1983) "Service providers in psychology. Results of the 1982 APA human resources survey." American Psychologist 36 (December): 1330-1352.

————and R. J. KILBURG (1981) "Health service providers in psychology. Results of the 1978 APA human resources survey." American Psychologist 36, 11: 1395-1418.

VAN VALEY, T. L., K. A. WOODS, and W. G. MARSTON (1982) "Patterns of segregation among Hispanic Americans: a base line for comparison." California Sociologist 5: 27-40.

VELEZ, W. (1984) "Finishing college: the effects of college type." Sociology of Education.

————(1983) The College Attainment Process in the United States During the Seventies. Ph.D. dissertation, Yale University.

VERNON, S. W. and E. ROBERTS (1982) "Use of the SADS-RDC in a tri-ethnic community survey." Archives of General Psychiatry 39: 47-52.

VEROFF, J., R. A. KULKA, and E. DONVAN (1981) Mental Health in America. New York: Basic Books.

WACHTER, M. and W. WASCHER (1983) "Labor market policies in response to structural changes in labor demand." Economics Department, University of Pennsylvania. (mimeo)

WAGLEY, C. and M. HARRIS (1958) Minorities in the New World. New York: Columbia University Press.

WALDINGER, R. (1982) "Immigrant enterprise and labor market structure." Cambridge, MA: Joint Center for Cuban Studies. (unpublished)

WARD, D. (1971) Cities and Immigrants. New York: Oxford University Press.

WARHEIT, G. J., C. E. HOLZER, III, and J. J. SCHWAB (1973) "An analysis of social class and racial differences in depressive symtomatology: a community study," Journal of Health and Social Behavior 14: 291-299.

WARHEIT, G. J., C. E. HOLZER, III, and S. A. AREY (1975) "Race and mental illness: an epidemiologic update." Journal of Health and Social Behavior 16: 243-256.

WARNER, S. (1971) The Private City: Philadelphia in Three Periods of Growth. Philadelphia: University of Pennsylvania Press.

WAYNE, L. (1984) "A pioneer spirit sweeps business," New York Times (March 25).

WEINBERG, M. (1983) The Search for Quality Integrated Education. Westport, CT: Greenwood.

WEISS, N. J. (1968) Charles Francis Murphy, 1858-1924: Respectability and Responsibility in Tammany Politics. Northhampton, RI: Smith College.

WEISSMAN, M. and J. K. MYERS (1978a) "Affective disorders in a U.S. urban community: the use of research diagnostic criteria in an epidemiological survey." Archives of General Psychiatry 35: 1304-1311.

——(1978b) "Rates and risks of depressive symptoms in a United States urban community." Acta Psychiatrica Scandinavica 57: 219-231.

——and P. S. HARDING (1978) "Psychiatric disorders in a U.S. urban community: 1975-76." American Journal of Psychiatry 135: 459-462.

WHITEHEAD, R., Jr. (1983) "The Chicago story: two dailies, a campaign and an earthquake." Columbia Journalism Review (July/August): 25-31.

——(1977) "The organization man." American Scholar 46: 351-357.

WHITMER, G. E. (1980) "From hospitals to jails: the fact of California's deinstitutionalized mentally ill." American Journal of Orthopsychiatry 50, 1: 65-75.

WILLIE, C. V., B. M. KRAMER, and B. S. BROWN (1973) Racism and Mental Health: Essays. Pittsburgh: University of Pittsburgh Press.

WILSON, J. Q. (1975) "Lock 'em up; and other thoughts on crime." New York Times Magazine (March 9): 11, 44-45.

——(1966) The Amateur Democrat: Club Politics in Three Cities. Chicago: University of Chicago Press.

WILSON, K. L. and A. PORTES (1980) "Immigrant enclaves: an analysis of the labor market experiences of Cubans in Miami." American Journal of Sociology 86 (September): 295-319.

WILSON, W. (1978) The Declining Significance of Race. Chicago: University of Chicago Press.

WINDLE, C., J. NEAL, and H. C. ZINN (1979) "Stimulating equity of services to nonwhites in community mental health centers." Community Mental Health Journal 15: 155-166.

WINSLOW, W. W. (1979) "The changing role of psychiatrists in community mental health centers." American Journal of Psychiatry 136: 24-27.

WITKIN, M. J. (1980) Trends in Patient Care Episodes in Mental Health Facilities, 1955-1977. Statistical Note 154. DHHS publication (ADM) 80-158. Rockville, MD: National Institute of Mental Health.

WILLIAMS, D. H., E. C. BELLIS, and S. W. WELLINGTON (1980) "Deinstitutionalization and social policy: historical perspectives and present delimmas." American Journal of Orthopsychiatry 50, 1: 54-64.

WIRTH, L. (1945) The Science of Man in the World Crisis. New York: Columbia University Press.

WOLFINGER, R. E. (1972) "Why political machines have not withered away and other revisionist thoughts." Journal of Politics 34: 365-398.

WOLKON, G. H., et al. (1973) "Ethnicity and social class in the delivery of services: Analysis of a child guidance clinic." American Journal of Public Health 64: 709-712.

WONG, C. L. (1983) "The implications of the 1980 Refugee Act for Southeast Asian refugees," in Special Services for Groups, Asian American Community Mental Health Training Center, Bridging Cultures: Southeast Asian Refugees in America.

WONG, M. G. (1983) "Chinese sweatshops in the United States: A look at the garment industry," in I. H. Simpson and R. L. Simpson (eds.) Research in the Sociology of Work: Volume II. Greenwich, CT: JAI.

———(1982) "The cost of being Chinese, Japanese and Filipino in the United States: 1960, 1970 and 1976." Pacific Sociological Review 25: 59-78.

———(1980) "Changes in socioeconomic status of the Chinese male population in the U.S. from the 1960-1970." International Migration Review 14: 511-524.

———and C. HIRSCHMAN (1983a) "The new Asian immigrants," in W. C. McCready (ed.) Culture, Ethnicity and Identity: Current Issues in Research. New York: Academic Press.

———(1983b) "Labor force participation and socioeconomic attainment of Asian-American women," Sociological Perspectives 26: 423-446.

WOODROOFE, K. (1971) From Charity to Social Work in England and the United States. Toronto, Canada: University of Toronto Press.

WOODS, R. A. (1903) Americans in Process. Boston: Houghton Mifflin.

WU, I. H. and C. WINDLE (1980) "Ethnic specificity in the relative minority use and staff of community mental health centers." Community Mental Health Journal 16: 156-168.

YAMAMOTO, J., O. C. JAMES, and N. PALLEY (1968) "Cultural problems in psychiatric therapy." Archives of General Psychiatry 19: 45-49.

YANCEY, W. L., L. RIGSBY, and J. D. McCARTHY (1972) "Social position and self-evaluation: The relative importance of race." American Journal of Sociology 78: 338-359.

About the Contributors

MARGO CONK is Associate Professor of History and Urban Affairs and Coordinator of Women's Studies at the University of Wisconsin—Milwaukee. She is the author of *U.S. Census* and *Labor Force Change* and numerous articles on the demographic history of the United States. She is currently working on a history of the census.

STEVEN P. ERIE is Associate Professor of Political Science at the University of California, San Diego. He worked as a policy analyst for the U.S. Department of Health and Human Services in 1980-1981. His research is on ethnic and minority group politics, urban political institutions, and federal social policy. He is currently completing a book on Irish big-city machines: *The Lost Hurrah: Irish-Americans and the Dilemma of Urban Machine Politics.*

RICARDO R. FERNANDEZ is Associate Professor in the Department of Cultural Foundations of Education at the University of Wisconsin—Milwaukee. Since 1977 he has served as Director of the Midwest National Origins Desegregation Assistance Center, which works with public school districts in ten states in meeting the educational needs of language minority students.

MARCIA FREEDMAN is Senior Research Scholar at the Conservation of Human Resources Project, Columbia University. She is a specialist in labor market operations as they affect employment and training. Her most recent work has been on new immigrants in urban labor markets, with a focus on New York City. She authored *Labor Markets: Segments and Shelters* (1976) and coauthored (with Anna Dutka) *Training Information for Policy Guidance* (1981).

PHILIP GARCIA is Visiting Assistant Professor of Chicano Studies at the University of California, Santa Barbara. He has published on labor

302

force inequalities among Mexican Americans. He is at work on a study of political orientations among Mexican Americans in the Southwest.

JOSE HERNANDEZ is currently Professor of Black and Puerto Rican Studies at Hunter College, City University of New York. He has also served as research director at the Latino Institute in Chicago in 1983. His research with census microdata dates to 1975 as Director of the Social Indicator Project at the U.S. Commission on Civil Rights, including consultation with the Bureau of the Census as member and Chair of the Hispanic Advisory Committee for the 1980 census.

LYNNE H. KLEINMAN is a doctoral student in the Urban Social Institutions program at the University of Wisconsin—Milwaukee. Her major research interests are in the area of urban history.

WILLIAM T. LIU is Professor of Sociology and Director of the Pacific/Asian American Mental Health Research Center at the University of Illinois at Chicago. He has published and authored books on the family, fertility, China, and the Philippines based on more than a decade of research in the Far East. In 1977-1978 he served as a panel staff member of the Board of the Public Committee on Mental Health.

ROBERT S. MAGILL is Associate Professor in the School of Social Welfare at the University of Wisconsin—Milwaukee. He has written *Community Decision Making for Social Welfare: Federalism, City Government and the Poor* and *Social Policy in American Society*.

LIONEL MALDONADO is Associate Professor of Sociology at the University of Wisconsin—Parkside. He has coauthored a monograph, *Chicanos in Utah* (1978), has written a chapter in *The Minority Report* (1982), and published a number of articles in social science journals on Chicanos in the United States.

ELEANOR M. MILLER is Assistant Professor of Sociology at the University of Wisconsin—Milwaukee. She has published work on trends in the criminality of women and has just completed a monograph on street hustling among underclass women.

JOAN MOORE is Professor of Sociology at the University of Wisconsin—Milwaukee. Most recently she has written (with Harry

Pachon) *Hispanics in the United States, Social Problems* (with Burton Moore), and *Homeboys: Gangs, Drugs and Prison in the Barrios of Los Angeles,* with four contributing authors from a community-based research group.

WILLIAM VELEZ is Assistant Professor of Sociology at the University of Wisconsin—Milwaukee. His major research interests are in the college attainment process and labor force experiences of Hispanics. His article "Finishing College: The Effects of College Type" recently appeared in the journal *Sociology of Education.*

MORRISON G. WONG is Assistant Professor of Sociology at Texas Christian University. His current research interests are Asian immigration, socioeconomic stratification, and social inequality. His publications have focused primarily on the Asian experience in America, examining such aspects as socioeconomic status and achievement, model student stereotypes, discrimination against Asians, female labor force participation, the elderly, regional variations, Chinese sweatshops, and Asian immigration.

ELENA S.H. YU is Associate Professor of Sociology and Psychiatry and Research Associate of the Pacific/Asian American Mental Health Research Center at the University of Illinois at Chicago. Her work includes a monograph, *Fertility and Kinship in Central Philippines* (with William T. Liu). She has contributed widely to social science journals and health and biomedical research publications. Her research includes family studies in China, psychiatric epidemiological surveys in Shanghai, refugee studies in San Diego, and fertility research in the Philippines.